Tracer Technology

FLUID MECHANICS AND ITS APPLICATIONS
Volume 96

Series Editor: R. MOREAU
MADYLAM
Ecole Nationale Supérieure d'Hydraulique de Grenoble
Boîte Postale 95
38402 Saint Martin d'Hères Cedex, France

Aims and Scope of the Series

The purpose of this series is to focus on subjects in which fluid mechanics plays a fundamental role.

As well as the more traditional applications of aeronautics, hydraulics, heat and mass transfer etc., books will be published dealing with topics which are currently in a state of rapid development, such as turbulence, suspensions and multiphase fluids, super and hypersonic flows and numerical modeling techniques.

It is a widely held view that it is the interdisciplinary subjects that will receive intense scientific attention, bringing them to the forefront of technological advancement. Fluids have the ability to transport matter and its properties as well as to transmit force, therefore fluid mechanics is a subject that is particularly open to cross fertilization with other sciences and disciplines of engineering. The subject of fluid mechanics will be highly relevant in domains such as chemical, metallurgical, biological and ecological engineering. This series is particularly open to such new multidisciplinary domains.

The median level of presentation is the first year graduate student. Some texts are monographs defining the current state of a field; others are accessible to final year undergraduates; but essentially the emphasis is on readability and clarity.

For further volumes:
http://www.springer.com/series/5980

Octave Levenspiel

Tracer Technology

Modeling the Flow of Fluids

 Springer

Octave Levenspiel
Chemical Engineering Department
Gleeson Hall, Oregon State University
Corvallis, OR 97331–2702, USA
levenspo@peak.org

ISSN 0926-5112
ISBN 978-1-4899-9211-6   ISBN 978-1-4419-8074-8 (eBook)
DOI 10.1007/978-1-4419-8074-8
Springer New York Dordrecht Heidelberg London

Printed on acid-free paper

Springer is part of Springer Science+Business Media (www.springer.com)

Preface

To tell how a vessel will behave as a heat exchanger, absorber, reactor, or other process unit, we need to know how fluid flows through the vessel.

In early engineering practice, the designer assumed either *plug flow* or *mixed flow* of the fluid through the vessel. But these idealized assumptions are often not good, sometimes giving a volume wrong by a factor of 100 or more, or producing the wrong product in multiple reactions systems.

The *tracer method* was then introduced, first to measure the actual flow of fluid through a vessel, and then to develop a suitable model to represent this flow. Such models are used to follow the flow of fluid in chemical reactors and other process units, in rivers and streams, and through solid and porous structures. Also, in medicine they are used to study the flow of chemicals, harmful or not, in the blood streams of animals and man.

This book shows how we use tracers to follow the flow of fluids and then we develop a variety of models to represent these flows. This activity is called Tracer Technology.

Corvallis, OR, USA   Octave Levenspiel

Contents

About the Author

Octave Levenspiel is Emeritus Professor of Chemical Engineering at Oregon State University, with primary interests in the design of chemical reactors. He was born in Shanghai, China in 1926, where he attended a German grade school, an English high school and a French Jesuit university. He started out to be an astronomer, but that was not in the stars, and he somehow found himself in chemical engineering. He studied at U. C. Berkeley and at Oregon State where he received his Ph.D. in 1952.

His pioneering book, "Chemical Reaction Engineering" was the very first in the field, has numerous foreign editions and has been translated into 13 foreign languages. His other books are, "The Chemical Reactor Omnibook", "Fluidization Engineering" (with co-author D. Kunii), "Engineering Flow and Heat Exchange", and "Understanding Engineering Thermo". He recently gathered his notes and musings together and self-published "Rambling Through Science and Technology".

He has received major awards from A.I.Ch.E. and A.S.E.E., and three honorary doctorates, from Nancy, France; from Belgrade, Serbia; from the Colorado School of Mines; and he has been elected to the National Academy of Engineering. Of his numerous writings and research papers, two have been selected as Citation Classics by the Institute of Scientific Information. But what pleases him most is being called the "Doctor Seuss" of chemical engineering.

Chapter 1
The Tracer Method

To tell how a vessel will behave as a heat exchanger, absorber, reactor, or other process unit, we need to know how the materials flow through the vessel. To do this exactly we have to measure the fluctuating velocity of flowing material at all points in the vessel and then analyze the results. This is an impractically complex procedure (Fig. 1.1).

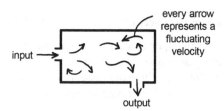

Fig. 1.1 Flow through real vessels

In early engineering practice, the designer assumed either *plug flow* or *mixed flow* of the fluid to represent the real vessel (Fig. 1.2).

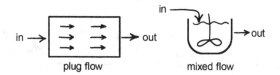

Fig. 1.2 The two simplest flow models

But these idealized assumptions are often not that good, so the *tracer method* was introduced to give better predictions for design.

In the tracer method, we introduce a tracer into the entering flow stream which exactly follows the flowing fluid (for example, blue ink into water).

O. Levenspiel, *Tracer Technology*, Fluid Mechanics and Its Applications 96, DOI 10.1007/978-1-4419-8074-8_1, © Springer Science+Business Media, LLC 2012

We then measure when the tracer leaves the vessel, and then analyze and interpret the results.

For linear processes, such as what happens in heat exchangers, absorbers, etc., and reactors processing first-order chemical reactions, this information is all that is needed to properly represent the behavior of the unit – no need to study the fluctuating velocity field within the vessel.

For second-order reactions and other nonlinear reactions, this procedure sometimes gives a reasonable approximation to the expected behavior (Fig. 1.3).

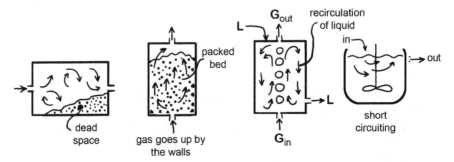

Fig. 1.3 Same of todays flow models

The questions we ask of the tracer experiments are the following:

- Are there any dead, stagnant, or unused regions in the vessel?
- Is there any channeling or bypassing of fluid in the vessel?
- Is there any circulation of fluid within the vessel or out of and back into the vessel?
- Can we develop a reasonable flow model to represent the flow?

To answer these questions we introduce tracer in one of a number of ways

- As a pulse (drop a glassful of red wine into flowing water)
- As a step input (switch from hot to cold water)
- As a periodic input
- As a random input (Fig. 1.4).

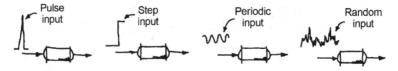

Fig. 1.4 Ideal tracer injection methods

The results from the pulse and step experiments are simpler to analyze and interpret while the periodic or random are much more difficult. So, in this volume, we focus on the pulse and step experiment.

We start with a pulse experiment. Here, we introduce a pulse of tracer into the fluid entering the vessel, and record when it leaves (Fig. 1.5).

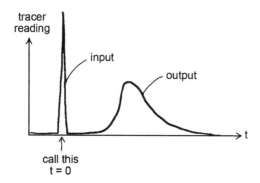

Fig. 1.5 Typical output to a pulse input

Two properties of the output curve are especially useful, the mean, $\bar{t}$, which tells on average when tracer leaves the vessel, and the variance, σ^2, which tells how broad the output curve is.

We characterize the flow in terms of these two quantities because they tell us what we want to know about the flow and they help us come up with useful flow models.

Fig. 1.3 ...

Chapter 2
The Mean and Variance of a Tracer Curve

Consider a vessel of volume V [m^3] and a volumetric flow rate v [m^3/s] of fluid in and out of the vessel. Then the space–time of the fluid in the vessel (a theoretical measure) is

$$\tau = \frac{V}{v} = \left[\frac{m^3}{m^3/s}\right] = [s]. \tag{2.1}$$

Now, let us do an experiment to study the actual flow of fluid inside and through the vessel. For this, at time $t = 0$, we introduce an instantaneous pulse of tracer into the fluid entering the vessel and we measure the mean and variance of the response curve of the fluid leaving the vessel. We call this the *pulse–response curve*. These quantities, $\bar{t}$ and σ^2, are widely used to characterize tracer curves, are simple to evaluate, but are often evaluated incorrectly, hence this presentation.

2.1 $\bar{t}$ and σ^2 from Experimental Pulse–Response Data

The two most useful measures for describing tracer curves, which are used in all areas of tracer experimentation, are

The mean, $\bar{t}$: tells when a tracer curve passes a measuring point, thus locating its "center of gravity" in time.

The variance, σ^2: tells how spread out in time, or how "fat" the curve is.

Given a smooth continuous pulse–response curve, C vs. t:

$$\text{the area under the tracer curve: } A = \int_0^\infty C \, dt, \tag{2.2}$$

$$\text{the mean of the curve: } \bar{t} = \mu = \frac{\int_0^\infty tC \, dt}{\int_0^\infty C \, dt}, \tag{2.3}$$

O. Levenspiel, *Tracer Technology*, Fluid Mechanics and Its Applications 96, DOI 10.1007/978-1-4419-8074-8_2, © Springer Science+Business Media, LLC 2012

the variance of the curve: $\sigma^2 = \dfrac{\int_0^\infty t^2 C\,dt}{\int_0^\infty C\,dt} - \bar{t}^2.$ \hfill (2.4)

The mean $\bar{t}$, or location of a simple triangular tracer curve is shown in Fig. 2.1.
And here are some quick estimates of the mean and variance if the curve looks
roughly rectangular, triangular, or nicely symmetrical and bell shaped (Fig. 2.2).

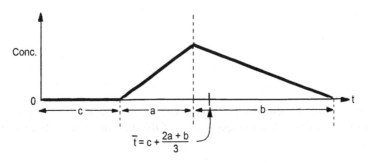

$$\bar{t} = c + \frac{2a+b}{3}$$

Fig. 2.1 Location of the mean of a triangle tracer curve

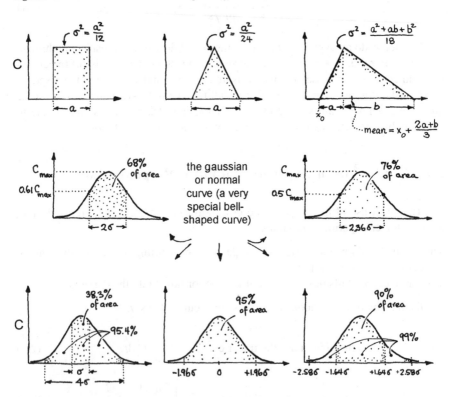

Fig. 2.2 Variance of some commonly met tracer curves

How you proceed depends on how much data and what kind of data are available. Did you take *instantaneous readings* (measure C and t at various times), or did you take *mixing cup measurements* (collect fluid for a minute every 5 min, and then measure its average concentration in each minute).

1. *Analysis of abundant data.* If the data, either mixing cup or instantaneous, are numerous and all closely spaced, we can estimate $\bar{t}$ and σ^2 by

$$\bar{t} = \frac{\sum t_i C_i \Delta t_i}{\sum C_i \Delta t_i} \text{ for equal time intervals } \frac{}{\Delta t_i} = \frac{\sum t_i C_i}{\sum C_i}, \qquad (2.5)$$

$$\sigma^2 = \frac{\sum t_i^2 C_i \Delta t_i}{\sum C_i \Delta t_i} - \bar{t}^2 \text{ for equal } \frac{\sum t_i^2 C_i}{\Delta t_i} \frac{}{\sum C_i} = \bar{t}^2. \qquad (2.6)$$

Try to use the same time interval Δt_i throughout.

2. *For a few data points* (instantaneous readings) draw a smooth curve through the data, then pick a number of points on the smoothed curve and evaluate $\bar{t}$ and σ^2 (Fig. 2.3).

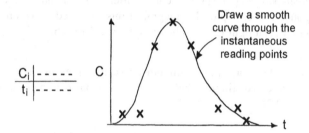

Fig. 2.3 Draw a curve for the instantaneous readings

3. *For a few mixing cup readings*, for example from $t = 1$ to 2, 3 to 4, 5 to 6,..., points A, B, C,..., draw the best estimate of the tracer curve points w, x, y,... Then draw a smooth curve through w, x, y,..., and evaluate $\bar{t}$ and σ^2 (Fig. 2.4).

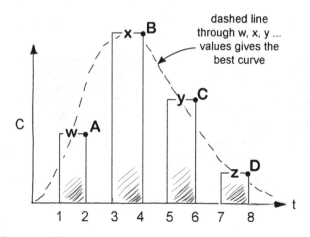

Fig. 2.4 Mixing cup readings w,x,y...

2.2 $\bar{t}$ and σ^2 from Step–Response Data

Consider a step experiment where the tracer goes from $C = 0$ to C_{max}.

The pulse–mean and the pulse–variance can be found from the corresponding step–response curve in a number of ways.

1. From the basic equation

$$\bar{t} = \frac{\int_0^\infty (C_{max} - C)\, dt}{C_{max}} \tag{2.7}$$

and

$$\sigma^2 = \frac{2 \int_0^\infty t(C_{max} - C)\, dt}{C_{max}} - \bar{t}^2. \tag{2.8}$$

2. For symmetrical S-shaped data
   Smooth S-shaped step–response curve often corresponds to a Gaussian pulse response curve. A plot the integral of this S-shaped curve on probability paper will give a straight line, as shown in Figs. 2.5 or 2.6.

3. For discrete data
   From the discrete data, draw a smooth curve to represent the step–response curve, taking care to distinguish between instantaneous readings and mixing cup readings (Fig. 2.7).

$$\bar{t} = \frac{\displaystyle\sum_{i-1}^{n} (C_{max} - C_i)\, \overset{\displaystyle t_i-t_{i-1}}{\Delta t_i}}{C_{max}} \quad \underset{=}{\text{or}} \quad \frac{\displaystyle\sum_{i=1}^{n} t_i \overset{\displaystyle C_i-C_{i-1}}{\Delta C_i}}{C_{max}} \tag{2.9}$$

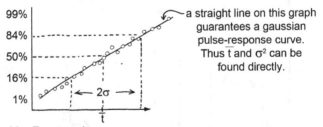

a straight line on this graph guarantees a gaussian pulse-response curve. Thus $\bar{t}$ and σ^2 can be found directly.

Fig. 2.5 Semi log F vs. t graph

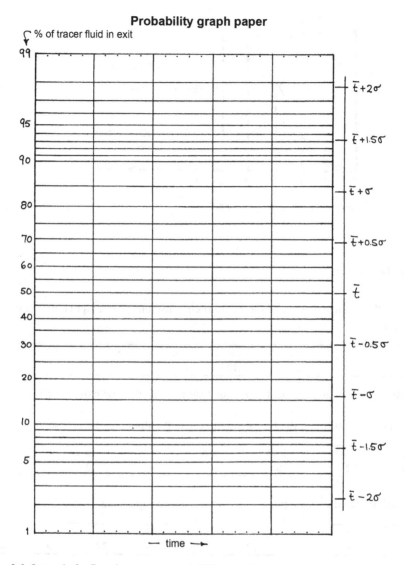

Fig. 2.6 Integral of a Gaussian curve on probability coordinates

At numerous equally spaced t_i values, tabulate the corresponding C_i value. Then from (2.7) and (2.8), or (2.9) and (2.10) evaluate $\bar{t}$ and σ^2

$$\sigma^2 = \frac{2\sum_{i=1}^{n} t_i(C_{\max} - C_i)\Delta t_i}{C_{\max}} - \bar{t}^2 \quad \text{or} \quad \frac{\sum_{i=1}^{n} t_i^2 \Delta C_i}{C_{\max}} - \bar{t}^2. \tag{2.10}$$

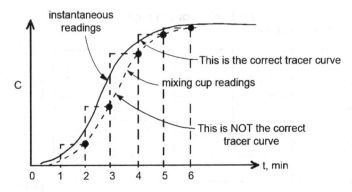

Fig. 2.7 Difference in mixing cup and instantaneous readings

Problems

1. Find the mean and variance for the following instantaneous output data to a pulse input.

t	0	2	4	6	8	10	12
C	0	2	10	8	4	2	0

The concentrations are instantaneous readings taken at the exit of the vessel.

2. Find the $\bar{t}$ and σ^2 for the following mixing cup output data to a pulse input.

t	0–2	2–4	4–6	6–8	8–10	10–12
C	2	10	8	4	2	0

3. Find the $\bar{t}$ and σ^2 for a pulse input from the step-input data.

t	<1	1	2	3	4	5	6	>6
C	10	10	22	26	28	29	30	30

Chapter 3
The E and E_θ Curves from Pulse and Step Tracer Experiments

3.1 The Pulse Experiment and the E Curve

At $t = 0$ [s], inject an instantaneous pulse of tracer **M** [kg] into fluid as it flows at a rate v [m³/s] in and out of a vessel of volume V [m³]. If evenly distributed in the vessel, this initial tracer concentration would be $C_0 = \mathbf{M}/V$ [kg/m³].

Tracer leaves the vessel at various times giving the C–t curve. All this is shown in Fig. 3.1.

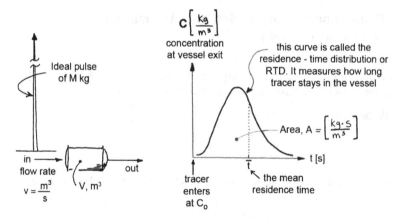

Fig. 3.1 The useful information obtained from pulse experiments

$$\begin{pmatrix} \text{Total area under the} \\ C \text{ curve} \end{pmatrix} : A = \int_0^\infty C \, dt \cong \sum_i \bar{C}_i \Delta t_i = \frac{\mathbf{M}}{v} \left[\frac{\text{kg} \cdot \text{s}}{\text{m}^3} \right], \qquad (3.1)$$

$$\begin{pmatrix} \text{Mean of the} \\ C \text{ curve} \end{pmatrix} : \bar{t} = \frac{\int_0^\infty tC \, dt}{\int_0^\infty C \, dt} \simeq \frac{\sum_i t_i \bar{C}_i \Delta t_i}{\sum_i \bar{C}_i \Delta t_i} = \frac{V}{v} \overset{\text{for equal}}{\underset{\Delta t_i}{=}} \frac{\sum_i t_i \bar{C}_i}{\sum \bar{C}_i} \, [\text{s}]. \qquad (3.2)$$

O. Levenspiel, *Tracer Technology*, Fluid Mechanics and Its Applications 96,
DOI 10.1007/978-1-4419-8074-8_3, © Springer Science+Business Media, LLC 2012

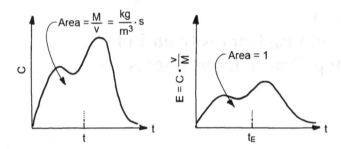

Fig. 3.2 Transforming an experimental C curve into an E curve

To find the **E** *curve* from the C curve, simply change the concentration scale such that the area under the C curve is unity. Thus, simply divide the concentration readings by **M**/v as shown in Fig. 3.2

$$\mathbf{E} = \frac{C}{\mathbf{M}/v} = \frac{\text{kg/m}^3}{\text{kg}/\,(\text{m}^3/\text{s})} = \left[\text{s}^{-1}\right]. \tag{3.3}$$

3.2   Dead Spaces, Stagnancies, and Adsorption of Tracer on Surfaces

One warning – the above relationship between the C and the **E** curve only holds exactly for vessels that have no internal dead spaces.

- With the whole vessel active

$$\bar{t} = \tau = \frac{V}{v}$$

- For vessels with dead spaces (maybe a dead rat in the vessel) or stagnancy

$$V_{\text{active}} < V, \quad \frac{V_{\text{active}}}{V} < 1, \quad \bar{t} < \tau.$$

- If $\bar{t} > \tau$ then either you have made some measurement error, or the tracer passes the measuring point more than once (this can happen when some fluid diffuses back to the measuring point and is measured more than once).
- Note that some curves have long, difficult to measure tails. In these cases, we may terminate our measurements at some t value. I often use about $3\bar{t}$ or $4\bar{t}$ as the cutoff point. Any fluid that stays longer than that time I will consider to be stagnant. So, with some of the curve unused we have (Fig. 3.3)

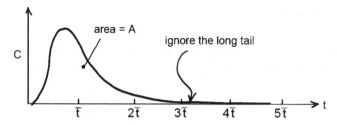

Fig. 3.3 Some C curves have long and difficult-to-measure tails

$$\bar{t} < \tau = \frac{V}{v}.$$

When there are no dead spaces in the vessel we can go one step further, by measuring time in terms of the number of mean residence times, or $\theta = t/\bar{t}$. Thus, with (3.3) and (3.4)

$$\mathbf{E}_\theta = \bar{t}\mathbf{E} = \bar{t}\frac{C}{A}, \ [-]. \tag{3.4}$$

Figure 3.4 shows how to transform $\mathbf{E}$ into $\mathbf{E}_\theta$

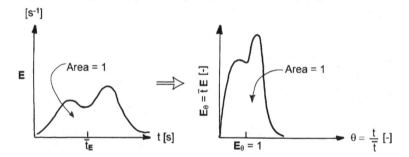

Fig. 3.4 This shows how to transform an $\mathbf{E}$ curve into an $\mathbf{E}_\theta$ curve

Overall then, to summarize,

$$\mathbf{E} = \frac{C}{\int_0^\infty C \, dt} = \frac{C}{A} \left[\frac{\text{kg/m}^3}{\text{kg} \cdot \text{s/m}^3} \right] = \text{s}^{-1}, \tag{3.5}$$

$$\mathbf{E}_\theta = \bar{t}\,\mathbf{E} \ [-]. \tag{3.6}$$

The $\mathbf{E}_\theta$ and $\mathbf{E}$ curves are useful in modeling studies.

3.2.1 Comments

(a) If a known quantity of tracer **M** (kg) is injected, the area under the C vs. t response curve, see Fig. 3.1, is

$$A_{\text{theoretical}} = A_{\text{th}} = \mathbf{M}/v \left[\text{kg s/m}^3 \right].$$

If $A_{\text{measured}} < A_{\text{th}}$ then stagnancies or dead spaces exist in the vessel.

$$\text{Fraction of vessel which is active} = \frac{A_{\text{measured}}}{A_{\text{th}}},$$

$$\text{Fraction of vessel which is dead} = \frac{A_{\text{th}} - A_{\text{measured}}}{A_{\text{th}}}.$$

If $A_{\text{measured}} > A_{\text{th}}$ then some tracer is adsorbed on the vessel walls or in the porous packing.

(b) *Reactive or decaying tracer*. Suppose the tracer is radioactive or reacts away by some sort of first order chemical process, say absorption onto the solid in the vessel, or slow chemical breakdown with half life $t_{1/2}$. Then, simply correct the observed concentration readings upward to what it should be without decay (Fig. 3.5).

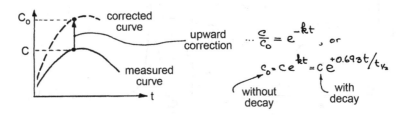

Fig. 3.5 Correcting adecaying tracer curve

If the time needed for significant decay of tracer is much longer than the time needed to make the tracer experiment then one need not consider making the upward correction. So if $t_{\text{decay}} \gg t_{\text{exp}}$ then $C_{\text{curve}} \cong C_{\text{correct}}$.

(c) *Warnings on Choice of Tracer and on the Experimental Method*

- Use a tracer with the same density as fluid, is not adsorbed, etc.
- Inject properly across the cross-section, proportional to the flow rate of the flowing fluid.
- Measure properly across the cross-section.

If you do not inject or measure the tracer properly you will not get a response which is the **E** curve. Chapter 9 deals with that problem (Fig. 3.6).

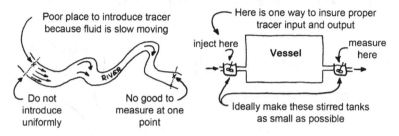

Fig. 3.6 Poor and good ways to use tracer in a vessel

3.3 The Step Experiment and the F Curve

Initially, v [m³/s] of fluid with no tracer flows through a vessel of volume V [m³]. Then, at time $t \geq 0$ switch the incoming flow to one with entering tracer $\dot{m}$ [kg/s] with unchanged flow rate v [m³/s]. A material balance for the tracer in the flowing fluid gives:

$$C_{max} = \frac{\dot{m}}{v} \left[\frac{kg}{m^3}\right] . \tag{3.7}$$

$$\text{Shaded area of Fig. 3.7} = C_{max}\bar{t} = \dot{m}\frac{V}{v^2} \left[\frac{kg \cdot s}{m^3}\right] . \tag{3.8}$$

This situation is shown in Fig. 3.7.

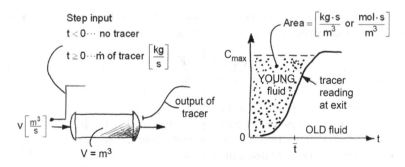

Fig. 3.7 Information obtained from a step tracer experiment

If the measured shaded area does not agree with and is smaller than the shaded area of Fig. 3.7, this means that dead spaces are present in the vessel.

If the measure of tracer concentration is changed so that the concentration starts at zero and rises to one, we get what is called the **F** curve, see Fig. 3.8.

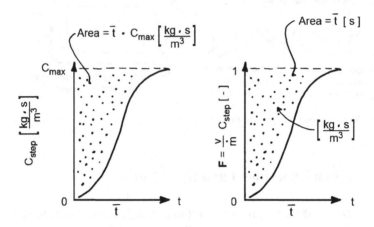

Fig. 3.8 Transforming an experimental C_{step} curve to an **F** curve

3.4   Relationship Between the E and F Curves

Since the step input is the integral of the pulse input, the output of the step is the integral of the pulse output. Thus at time t, with (3.4),

$$\frac{C_{step}}{C_{step, max}} = \mathbf{F} = \int_0^t \mathbf{E}\,dt = \left.\int_0^t C_{pulse}\,dt \middle/ \text{total area} \right. \tag{3.9}$$

On the other hand if you differentiate (take the slope of) the **F** curve at any time t you will get the corresponding **E** value at that time

$$\frac{_0\!\int C_{step}\,dt}{C_{step, max}} = \frac{d\mathbf{F}}{dt} = \mathbf{E} \tag{3.10}$$

These relationships are shown in Fig. 3.9.

From the **E** or **F** curves we can try to create flow models for the vessel. For this the **E** curve is usually more useful because an **F** curve which, if obtained from somewhat scattered data, is often quite useless.

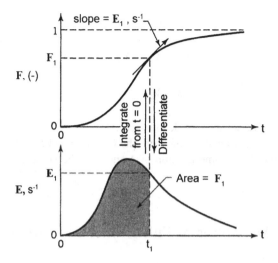

Fig. 3.9 Relationship between the **E** and **F** curves

3.5 Two or More Immiscible Flowing Streams (G/L, G/S, L/L, L/S, S/S)

Streams of two immiscible fluids or solids pass through a vessel of volume V [m^3] at flow rates v_A and v_B [m^3/s]. Find the vessel fraction which contains A V_A, which contains B V_B, and which is stagnant V_{dead}.

For this, introduce tracer into stream A, then into stream B, both at $t = 0$ and find their RTD's (Fig. 3.10).

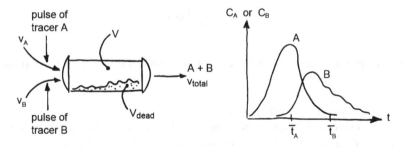

Fig. 3.10 Two immiscible inputs and their outputs

By material balance

$$\bar{t}_A = V_A/v_A, \tag{3.11}$$

$$\bar{t}_B = V_B/v_B, \tag{3.12}$$

$$V_A + V_B = V_{\text{active}} \tag{3.13}$$

$$V_{\text{total}} - V_{\text{active}} = V_{\text{dead}} \tag{3.14}$$

Special case. A vessel with no dead spaces.

In this case, one only needs to measure the flow rate and the RTD of one of the two phases, to find the volume fraction of A and B in the vessel.

3.6   Two or More Streams Either Enter or Leave the Vessel

Where the fluids are miscible but are not well mixed in the vessel, measure the flow rate of A and of B, and record the RTD of A. This gives $\bar{t}_A$. Then a material balance gives

$$\bar{t}_A = V_{\text{active}}/v_A \tag{3.15}$$

$$\bar{t}_B = V_{\text{active}}/v_B \tag{3.16}$$

$$v_A + v_B = v_{\text{total}} \tag{3.17}$$

$$\bar{t}_{\text{active}} = V_{\text{active}}/(v_A + v_B) \tag{3.18}$$

$$V_{\text{total}} = V_{\text{active}} + V_{\text{dead}} \tag{3.19}$$

$$\tau = \bar{t}_{\text{total}} = \frac{V_{\text{total}}}{v_{\text{total}}} \tag{3.20}$$

These equations give you the necessary information to tell what is happening in the vessel.

Example 1. *E Curves from Experiment.*
The instantaneous concentration readings and sketch below represent the response to a pulse input called a Dirac δ-function, to a vessel. Calculate and graph the corresponding **E** vs. t and $\mathbf{E}_\theta$ vs. θ curves (Fig. 3.11).

Time t (min)	0	5	10	15	20	25	30	35
Tracer output reading C (mol/L)	0	3	5	5	4	2	1	0

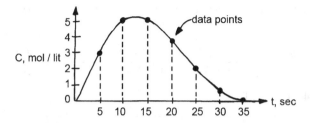

Fig. 3.11 Instantaneous experimental readings

Solution. To determine the **E** curves first estimate the area **A** under the *C* vs. *t* curve. For this draw rectangles (Fig. 3.12).

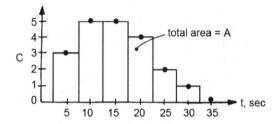

Fig. 3.12 Rectangles represent the instantaneous readings

Then from (3.1) determine the total area

$$A = \sum C\Delta t = [3 + 5 + 5 + 4 + 2 + 1]\, 5 = 100 \frac{\text{mol}\cdot\text{min}}{\text{L}}$$

From (3.3)

$$\mathbf{E} = \frac{C_{\text{pulse}}}{\text{total area}} = \frac{C}{100}\left[\frac{\text{mol/L}}{\text{mol}\cdot\text{min}/\text{L}} = \text{min}\right]$$

Thus, the data, see table below, and graph, Fig. 3.13, are

t [min]	0	5	10	15	20	25	30	35
E [min^{-1}]	0	0.03	0.05	0.05	0.04	0.02	0.01	0

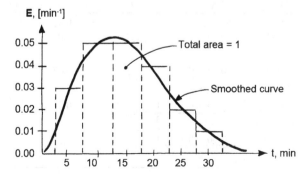

Fig. 3.13 Fitted **E** curve through the data

Next, to find $\mathbf{E}_\theta$ we must measure time in terms of $\bar{t}$, thus $\theta = t/\bar{t}$ and $\mathbf{E}_\theta = \bar{t}\mathbf{E}$. So from (3.2)

$$\bar{t} = \frac{\sum tC}{C} = \frac{5 \cdot 3 + 10 \cdot 5 + 15 \cdot 5 + 20 \cdot 4 + 25 \cdot 2 + 30 \cdot 1}{3 + 5 + 5 + 4 + 2 + 1} = \frac{300}{20} = 15 \text{ min}$$

and from (3.4) (Fig. 3.14)

$\theta = \dfrac{t}{15}$ [min]	0	$\dfrac{1}{3}$	$\dfrac{2}{3}$	1	$\dfrac{4}{3}$	$\dfrac{5}{3}$	2	$\dfrac{7}{3}$
$\mathbf{E}_\theta = 15\ \mathbf{E}$ [-]	0	0.45	0.75	0.75	0.60	0.30	0.15	0

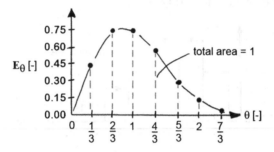

Fig. 3.14 $\mathbf{E}_\theta$ vs. θ curves

Example 2. *The F Curve from a Pulse Experiment.*
From the experimental data of Example 1 tabulate and plot the corresponding **F** function.

Solution. The **F** function in θ units is

$$\mathbf{F} = \int_0^\infty \mathbf{E}_\theta \, d\theta \xrightarrow{\begin{array}{c}\text{for discrete}\\\text{points}\end{array}} = \sum \mathbf{E}_\theta \Delta\theta$$

Table E2

i	θ	$\mathbf{E}_\theta$	Area present in the interval between $i-1$ and i, $\bar{\mathbf{E}}_i \Delta\theta_i = \frac{\mathbf{E}_i + \mathbf{E}_{i-1}}{2}(\theta_i - \theta_{i-1})$	Summing all θ_i $\mathbf{F} = \sum\limits^i \mathbf{E}_\theta \Delta\theta$
0	0	0	0	0
1	1/3	0.45	$\dfrac{0.45 + 0}{2} \times \dfrac{1}{3} = 0.075$	0.075
2	2/3	0.75	$\dfrac{0.75 + 0.45}{2} \times \dfrac{1}{3} = 0.200$	0.275
3	1	0.75	$\dfrac{0.75 + 0.75}{2} \times \dfrac{1}{3} = 0.250$	0.525
4	4/3	0.60	$\dfrac{0.60 + 0.75}{2} \times \dfrac{1}{3} = 0.225$	0.750

(continued)

Table E2 (continued)

i	θ	E_θ	Area present in the interval between $i-1$ and i, $\bar{E}_i\Delta\theta_i = \frac{E_i+E_{i-1}}{2}(\theta_i - \theta_{i-1})$	Summing all θ_i $F=\sum\limits^i E_\theta\Delta\theta$
5	5/3	0.30	$\dfrac{0.30+0.60}{2}\times\dfrac{1}{3}=0.150$	0.900
6	2	0.15	$\dfrac{0.15+0.30}{2}\times\dfrac{1}{3}=0.075$	0.975
7	7/3	0	$\dfrac{0+0.15}{2}\times\dfrac{1}{3}=0.025$	1.000

Table E2 shows the details of the summation which is done by the trapezoidal rule.

Summing gives $\sum_i E_\theta\Delta\theta = 1.00$

Graphically we show this (Fig. 3.15)

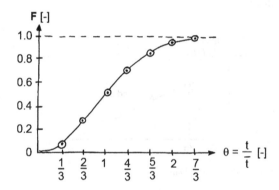

Fig. 3.15 F value from the last of Table EZ

Example 3. Pulse Modeling.

Pure water flows at $v = 4$ L/min through a vessel of volume V liter. At time $t = 0$ and thereafter, blue ink at $\dot{m} = 1$ g/min is fed to the entering fluid stream and its concentration in the exit stream is recorded, as shown in Fig. 3.16.

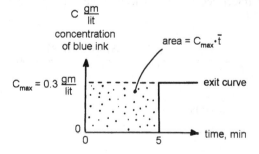

Fig. 3.16 Observed readings

What is the volume of the vessel?

Solution. By material balance, (3.7) and (3.8) give:

$$\bar{t} = 5 \text{ min, and } C_{max} = \frac{m}{v} = \frac{1 \text{ g/L}}{4 \text{ L/min}} = 0.25 \frac{g}{L},$$

from the above graph $C_{max} = 0.3 \frac{g}{L}$

These do not match. More tracer leaves than enters. So something is wrong!!
Redo the experiment.

Example 4. *Water Flow in the Columbia River.*

A batch of radioactive material is dumped into the Columbia River near Richmond, WA. At Bonneville Dam about 240 miles downstream, the flowing waters (6,000 m^3/s) are monitored for a particular radioisotope ($t_{1/2} > 10$ years), and the following data Fig. 3.17 are obtained.

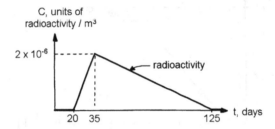

Fig. 3.17 Flow of river at Bonneville Dam

(a) How many units of this tracer were introduced into the river?
(b) What is the volume of Columbia River waters between Bonneville dam and the point of introduction of tracer?

Solution.

(a) For a 10 year tracer half life and with a much smaller time for the experiment we can ignore tracer decay. So the mean of the tracer curve, from geometry or from Fig. 3.17 is

$$\bar{t} = \frac{\sum tC}{\sum C} = \frac{30(15 \times 10^{-6}) + 65(90 \times 10^{-6})}{105 \times 10^{-6}} = 60 \text{ days.}$$

The area under the curve is

$$A = (125 - 20)(10^6) = 105 \times 10^{-6} \frac{\text{units day}}{\text{m}^3}.$$

From the material balance, (3.1), the amount of tracer introduced is

$$\mathbf{M} = A \cdot v = \left(105 \times 10^{-6} \frac{\text{unit day}}{\text{m}^3}\right)\left[\left(6000 \frac{\text{m}^3}{\text{s}}\right)\left(3,600 \times 24 \frac{\text{s}}{\text{day}}\right)\right]$$
$$= 54,432 \text{ units of tracer.} \leftarrow \text{(a)}$$

(b) Because $\bar{t} = V/v$, the volume of this section of the river is

$$V = \bar{t} \times v = (60 \text{ days})\left[\left(6,000 \frac{\text{m}^3}{\text{s}}\right)\left(3,600 \times 24 \frac{\text{s}}{\text{day}}\right)\right]$$
$$= 3.11 \times 10^{10} \text{ m}^3. \leftarrow \text{(b)}$$

Example 5. *Fraction of Gas and of Liquid in a Pipe.*
A pipeline (10 cm ID, 19.1 m long) simultaneously transports gas and liquid from here to there. The volumetric flow rate of gas and liquid are 60,000 and 300 cm³/s, respectively. Pulse tracer tests on the fluids flowing through the pipe give results as shown below. What fraction of the pipe is occupied by gas and what fraction by liquid? (Fig. 3.18).

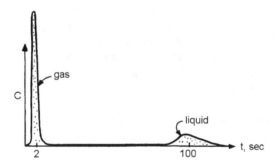

Fig. 3.18 Flow rate of G and L in a pipeline

Solution. Given the following flow rates and volumes

$$v_g = 60,000 \text{ cm}^3/\text{s},$$

$$v_l = 300 \text{ cm}^3/\text{s},$$

$$\text{Volume of pipe} = \frac{\pi}{4}d^2(\text{L}) = \frac{\pi}{4}(10)^2(1,910) = 150,011 \text{ cm}^3.$$

Volume occupied by gas V_g and by liquid V_ℓ, from (3.11) and (3.12), are

$$V_g = \bar{t}_g v_g = 2(60,000) = 120,000 \text{ cm}^3,$$

$$V_l = \bar{t}_l v_l = 100(300) = 30,000 \text{ cm}^3.$$

Therefore, the fraction of gas and of liquid are

$$\left.\begin{array}{l} \text{Frac}_g = \dfrac{V_g}{V_{\text{total}}} = \dfrac{120,000}{120,000 + 30,000} = 0.8 \\[2mm] \text{Frac}_l = \dfrac{30,000}{120,000 + 30,000} = 0.2 \end{array}\right\} \leftarrow$$

Problems

The following three graphs show the tracer output from the vessel for a pulse input at time $t \geq 0$. For each case, sketch the age distribution functions **E** and **E$_\theta$**, and find what else is required by the problem statement. To make the mathematics bearable, only simple idealized curves are shown (Figs. 3.19–3.21).

1.

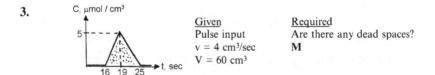

Given	Required
Pulse input	Are there any dead spaces?
M = 1 mol at t = 0 V	
v = 4 lit/min	

Fig. 3.19 Pulse tracer test

2.

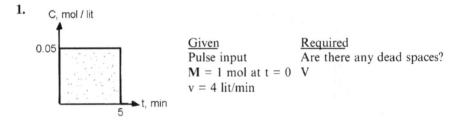

Given	Required
Pulse input	Are there any dead spaces?
M unknown	**M, v**
V = 500 lit	

Fig. 3.20 Pulse tracer test

3.

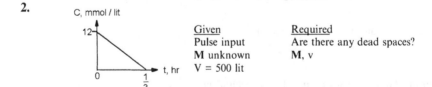

Given	Required
Pulse input	Are there any dead spaces?
v = 4 cm³/sec	**M**
V = 60 cm³	

Fig. 3.21 Pulse tracer test

The following three graphs show the tracer output from a step input. For each case tell, if possible, whether any dead space exists in the vessel and sketch the **F** curve, and the corresponding **E** and **E**$_\theta$ curves. Again, to make the mathematics bearable, only simple idealized curves are shown (Figs. 3.22–3.24).

4.

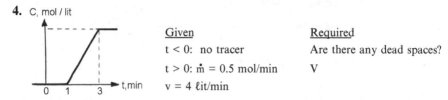

Given	Required
$t < 0$: no tracer	Are there any dead spaces?
$t > 0$: $\dot{m} = 0.5$ mol/min	V
$v = 4$ lit/min	

Fig. 3.22 Step output

5.

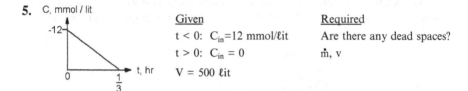

Given	Required
$t < 0$: $C_{in} = 12$ mmol/lit	Are there any dead spaces?
$t > 0$: $C_{in} = 0$	$\dot{m}$, v
$V = 500$ lit	

Fig. 3.23 Step output

6.

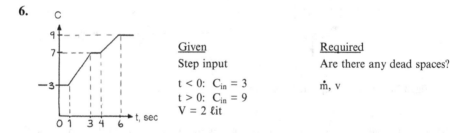

Given	Required
Step input	Are there any dead spaces?
$t < 0$: $C_{in} = 3$	$\dot{m}$, v
$t > 0$: $C_{in} = 9$	
$V = 2$ lit	

Fig. 3.24 Step output

7. On the average, 1 man/min and 2 women/min enter the department store. On the average, men shop for 1 h and women shop for 4 h. I wonder what fraction of customers in the store are men, what fraction are women?

8. A big stirred tank 300 m³ in volume is fed 24 m³ gas/min and 16 m³/min of a liquid–solid slurry. I worry that the stirrer may be underpowered and that some solid which might have settled to the bottom of the tank is not flowing.
To find out, I introduce pulses of tracer into the two feed streams with results as shown in Fig. 3.25. What fraction of the tank, if any, has a deposit of solid?

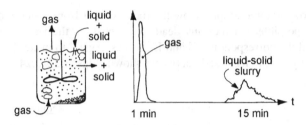

Fig. 3.25 Sketch and graph for problem 8

9. A bubble column with a 10 m³ void volume is fed with 10 m³ gas/min and 1 m³ liquid/min and its gas/liquid ratio is supposed to be about 50/50%. However, it is behaving badly and I suspect that the G/L volume ratio is not what I expect it to be.

   To check this I introduce a pulse of tracer into the entering gas stream and record when it leaves, see Fig. 3.26. What does it tell you about the vessel contents?

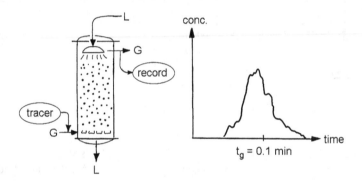

Fig. 3.26 Are solids deposited in the tank?

10. On the average, 2,000 men and 2,000 women enter this college of 15,460 students each fall term. After 1 year, 500 men and 200 women drop out, the rest go on to graduate. On the average, men need 4.8 years to graduate. How long does it take for women to graduate?

Chapter 4
Two Ideal Flow Models: Plug Flow and Mixed Flow

4.1 Plug Flow

Here, all the fluid passes through the vessel in single file, with no mixing of earlier with later entering fluid, no overtaking, or backmixing.

There are three versions of this model, one which does not allow lateral mixing of flowing elements, the second which does allow such mixing.

When we say plug flow we usually mean the first case. Other names for *plug flow* are

- Piston flow
- Ideal tubular flow
- PFR (plug flow reactor)
- ITR (ideal tubular reactor)

There is a third version of the term "plug flow." Here, the flowing elements mix with other older and younger fluid, but then at time $\bar{t}$ the elements unmix to leave the vessel. One would think that this would violate the second law of thermo and cannot happen in nature, but there are some situations where we can make it happen and it does occur.

The output **E** tracer curve for all three versions of plug flow is given by the unique mathematical expression, the Dirac delta function, $\delta(t - \bar{t})$, which represents a perfect pulse output at time $t = \bar{t}$, or $\delta(t - \bar{t}) = 0$. The output **F** curve is an ideal step function. These curves are shown in Figs. 4.1 and 4.2.

4.2 Mixed Flow

On entering the vessel the fluid mixes instantaneously and uniformly with the vessel's contents. In effect, every bit of fluid loses its memory as it enters the vessel and it could be anywhere in the vessel (Fig. 4.3).

O. Levenspiel, *Tracer Technology*, Fluid Mechanics and Its Applications 96, DOI 10.1007/978-1-4419-8074-8_4, © Springer Science+Business Media, LLC 2012

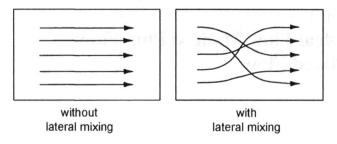

without
lateral mixing

with
lateral mixing

Fig. 4.1 Two forms of plug flow

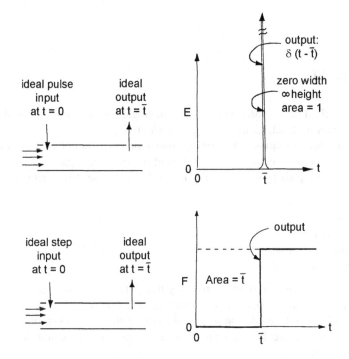

ideal pulse
input
at t = 0

ideal
output
at t = t̄

E

output:
δ (t - t̄)

zero width
∞ height
area = 1

ideal step
input
at t = 0

ideal
output
at t = t̄

F   Area = t̄

output

Fig. 4.2 Response curves for plug flow

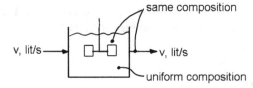

same composition

v, lit/s

v, lit/s

uniform composition

Fig. 4.3 Mixed flow

We call this type of flow *mixed flow*. Other terms sometimes used are:

- Completely mixed vessel
- Uniformly mixed vessel
- IST (ideal stirred tank)

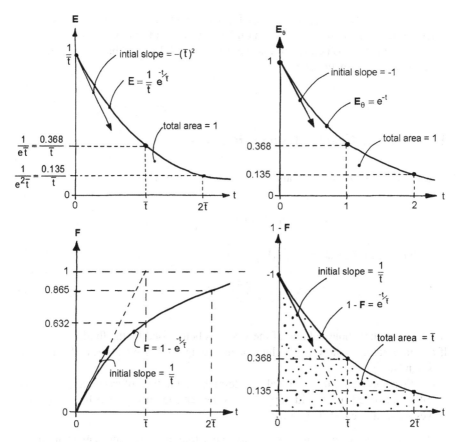

Fig. 4.4 Response curves for mixed flow

- CSTR (continuous stirred-tank reactor)
- CFSTR (constant flow stirred-tank reactor)
- C* (C star reactor)
- Perfectly mixed reactor
- MFR (mixed flow reactor)

The **E** and **F** curves for mixed flow are shown in Fig. 4.4.

4.3 Hybrid Plug-Mixed Model

Suppose fluid enters a vessel V, say a stirred tank, at a flow rate v, mixes with the vessel contents, but after staying in the vessel for time $\tau = \bar{t} = V/v$ each of the entering molecules separates from its neighbors, goes to the exit of the vessel, and leaves.

We also assume that $v_{\text{input}} = v_{\text{output}}$. Thus, every molecule stays for a time $\tau = \bar{t}$ in the vessel. This represents an even different form of plug flow.

This flow pattern is neither ordinary plug or mixed flow, but has aspects of both these extreme flow patterns.

4.4 Dead or Nearly Stagnant Regions

Measure the vessel volume V and the fluid flow rate v. Then

$$\text{space time} : \tau = \frac{V}{v}$$

and the mean of the tracer curve (Fig. 4.5)

$$\bar{t} = \frac{\int_0^\infty tC \, dt}{\int_0^\infty C \, dt}.$$

If $\bar{t} \cong \tau$ then the whole volume of the vessel is being used by the fluid.

If $\bar{t} < \tau$ then some of the vessel volume is not being used. We call these the dead or stagnant regions.

If $\bar{t} > \tau$ then something in the vessel is holding back the flow of tracer. The tracer is either adsorbed on a solid surface or is denser and sinks to the bottom of the vessel, or it is eaten by goldfish, etc.

The tracer curves in some vessels have very long and difficult to evaluate tails. In such cases, we often arbitrarily decide to cut off the tracer curve at some point. I often choose $3\bar{t}$ for the cutoff point (Fig. 4.6).

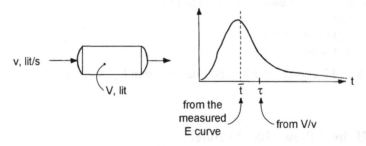

Fig. 4.5 Mixed flow with some stagnant regions

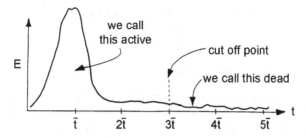

Fig. 4.6 For long tails we often use a cut off point

Example 1. *The Dirty Pipeline.*

Dirty waste water containing sticky black solid is discharged from a 111-m long 10.16-cm ID plastic pipe. In the cleaning cycle, the pipe is flushed with a steady stream of fresh water. The flow rate of water is 30 L/min, and a tracer test of the water in the pipe is shown below. Check whether the flow of fresh water is reasonably approximated by plug flow or by mixed flow and whether any of the black solid may have deposited on the pipe walls (Fig. 4.7).

Fig. 4.7 Switch from dirty to clean water

Solution. From the graph the fresh water flows in plug flow.

$$\text{Volume of clean pipeline} = \frac{\pi}{4}d^2L = \frac{\pi}{4}(10.16)^2(11,100) = 899,912 \text{ cm}^2$$
$$= \sim 900 \, L,$$

$$\text{Flow rate in clean pipeline: } v = 30 \, L/\min,$$

$$\left.\begin{array}{c} \text{Therefore the space time is } \tau = \dfrac{900}{30} = 30 \text{ min} \\[2mm] \text{From the tracer curve } \bar{t} = 27 \text{ min} \end{array}\right\}$$

So the fraction of deposit in the pipe

$$= \frac{30 - 27}{30} = 0.1 = 10\%.$$

Example 2. *The History of Dollar Bills.*

From the *New York Times Magazine*, December 25, 1955, comes the following noteworthy news item: "The United States Treasury reported that it costs eight-tenths of a cent to print a dollar bill, and that of the billion and a quarter now in circulation, a billion have to be replaced annually." Assume that the bills are put into circulation at a constant rate and continuously; that the bills are withdrawn from circulation without regard to their condition, in a random manner, and that there is no change in the total number of bills in circulation.

(a) Find the average age of the bills being replaced (or the bills in circulation).
(b) Determine the age distribution of a representative batch of bills being taken out of circulation (or in circulation).
(c) What fraction of bills is used for over 4 years?
(d) At any time, how many bills are over 21 years old in the exit stream, and are still in circulation?

Solution. The analogy with mixed flow should be evident. All we need to do is identify the various quantities. Referring to Fig. 4.8, we have:

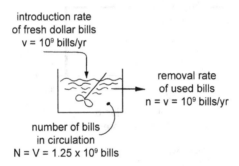

Fig. 4.8 The life of US dollar bills

(a) From this sketch or data, the mean residence *time*, or the mean life of a bill is

$$\bar{t} = \frac{V}{v} = \frac{1.25 \times 10^9 \text{ bills}}{10^9 \text{ bills/year}} = \underline{\underline{1.25 \text{ years.}}}$$

(b) The RTD of a representative batch of bills being removed, from Fig. 4.4 is

$$\mathbf{E} = \frac{1}{\bar{t}} e^{-t/\bar{t}} = 0.8 e^{-0.8t}$$

(c) Fraction of bills 4 years or older in the discard stream, or in circulation is

$$\int_4^\infty \mathbf{E}\, dt = \int_4^\infty 0.8e^{-0.8t}\, dt = 0.8 \times \frac{e^{-0.8t}}{0.8}\bigg|_4^\infty = -\frac{1}{e^\infty} + \frac{1}{e^{3.2}} = 4.08\%.$$

(d) First find the fraction of bills older than 21 years old (ancient bills) in the discard stream or in circulation.

$$\int_{21}^\infty \mathbf{E}\, dt = -\left[\frac{1}{e^\infty} - \frac{1}{e^{21 \times 0.8}}\right] = [0 + 5.0565 \times 10^{-8}].$$

So, in the exit stream the number of ancient bills are

$$(5.0565 \times 10^{-8})10^9 = \underline{\underline{50.6/\text{year.}}}$$

The number of ancient bills in use at that time are

$$(5.0565 \times 10^{-8})(1.25 \times 10^9) = \underline{63.2 \text{ bills.}}$$

Problems

1. Referring to Example 2, suppose a new series of dollar bills is put into circulation at a given instant in place of the original series.

   (a) What is the fraction of new bills in circulation or being withdrawn at any time?
   (b) Plot this equation.

2. Referring to Example 2 and Problem 1, suppose that during a workday a gang of counterfeiters put in circulation one million dollars in fake one dollar bills.

   (a) If not detected, what will be the fraction being withdrawn (or in circulation) at any time?
   (b) After 10 years how many of these bills are still in circulation?

3. The army kitchen company has to feed the whole Army division. For breakfast they have designed a large egg boiler to prepare 3-min soft boiled eggs. The boiler holds 180 eggs which are dropped in at a steady rate and removed at a steady rate. In the boiler, they are gently stirred and well mixed, and have a 3 min mean residence time. However, there have been complaints – some eggs over boiled, others under boiled.

Eggs which are boiled for more than 4 min are considered over boiled, less than 2 min, under boiled. What fraction of the eggs from this cooker are either under, or over boiled?

4. Perchlorate salts provide oxygen for solid fuels – rockets, missiles, roadside flares, automobile airbags, matches, etc. Kerr McGee Chemical Co. makes this material in a chemical plant near Las Vegas, and in the last 50 years this plant has been disposing its waste into a stream that flows into the Colorado River.

This water flows more than 600 km, over Hoover Dam, Parker Dam, Palo Verde Dam, Imperial Dam, and on to the Mexican border. Part of this water is diverted to the Colorado River Aqueduct, the All-American Canal, the Coachella Canal, the Imperial Dam, and the Central Arizona Project, and it is used by over 15 million people in Los Angeles, San Diego, Southern California, and Arizona.

Now, it has recently been discovered that perchlorates can affect the nervous systems of animals, and that it also disrupts the thyroid function of fetuses. This is a worry because analyses conducted in 2003 found perchlorates of about 4–5 ppb in lettuce irrigated with this water. This finding then makes it a nationwide concern because the Imperial Valley is the country's biggest producer of lettuce.

The EPA, in their draft assessment released in 2002, suggested a health protection standard of 1 ppb. The Pentagon objected strongly arguing for a higher benchmark, with some of its contractors recommending 200 ppb. The Bush administration has asked the NAS to review the data and suggest a drinking water standard [*Chem. Eng. News*, 37 (August 18, 2003)].

Until 1999, about 400–500 kg/day of perchlorate salts were dumped into the Colorado River. Suppose that the discharge was stopped completely on August 1, 1999. If on April 1, 2003, the perchlorate concentration in water at a downstream monitoring station was found to be 140 ppb, but 136 ppb 1 month later, when can we expect it to drop to 1 ppb at this measuring station? For more see *Chem. and Eng. News*, pg 25, October 13, 2008.

5. Some of our students take the biochem option, others a management option, the design option, and so on. However, ours is an outstanding engineering school because none of our students fail. They all graduate in 4 years and have good self-esteem. If a random batch of entering students is tracked, what flow model best represents their progress?

Chapter 5
Compartment Models

Flow models can be of different levels of sophistication, and the compartment models of this chapter are the next stage beyond the very simplest, those that assume the extremes of plug flow or mixed flow. In the compartment models, we consider the vessel and the flow through it as follows:

$$\text{Total volume } V \dots \begin{cases} V_p & - \text{ plug flow region} \\ V_m & - \text{ mixed flow region} \\ V_d & - \text{ dead or stagnant region within the vessel,} \end{cases} \left.\begin{array}{c} \\ \end{array}\right\} V_a \; - \text{ active volume} \tag{5.1}$$

$$\text{Total throughflow } v \dots \begin{cases} v_a & - \text{ active flow, that through the plug} \\ & \quad \text{and mixed flow regions,} \\ v_b & - \text{ bypass flow,} \\ v_r & - \text{ recycle flow,} \end{cases} \tag{5.2}$$

$$v = v_a + v_b, \tag{5.3}$$

$$\text{Space time: } \tau = \frac{V}{v}, \tag{5.4}$$

$$\text{Mean of the tracer curve: } \bar{t} = \frac{V_a}{v}, \tag{5.5}$$

$$\text{If there are no dead regions: } \tau = \frac{V}{v} = \frac{V_a}{v} = \bar{t} \tag{5.6}$$

5.1 Tracer Curves

By comparing the **E** curve for the real vessel with the theoretical curves for various combinations of compartments and throughflow, we can find which model best fits the real vessel. Of course, the fit may not be perfect; however, models of this kind are often a reasonable approximation to the real vessel.

O. Levenspiel, *Tracer Technology*, Fluid Mechanics and Its Applications 96, DOI 10.1007/978-1-4419-8074-8_5, © Springer Science+Business Media, LLC 2012

The next few pages show what the **E** curves and the **F** curves look like for various combinations of the above elements – certainly not all combinations (Figs. 5.1–5.3).

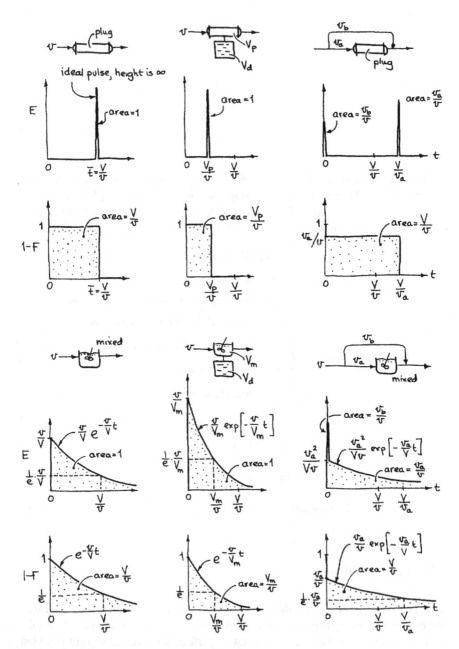

Fig. 5.1 E and F curves for various combination of vessels

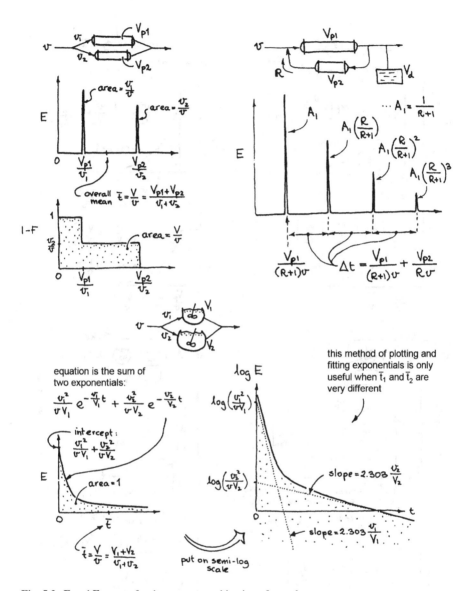

Fig. 5.2 E and F curves for three more combination of vessels

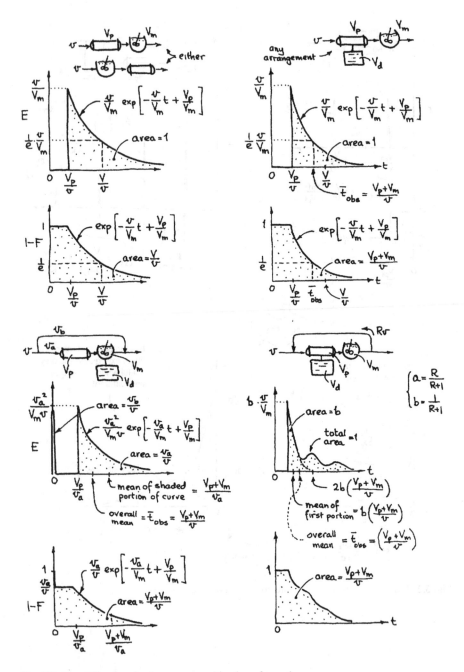

Fig. 5.3 E and F curves for four more combination of vessels

5.1.1 Hints, Suggestions, and Possible Applications

(a) We must know τ, thus both V and v, if we want to properly evaluate all the elements of these models, including dead spaces. If we only measure $\bar{t}_{obs}$ we cannot find the size of the stagnant regions and must ignore them in our model building. Thus,

If the real vessel has dead spaces: $\bar{t}_{obs} < \tau$
If the real vessel has no dead spaces: $\bar{t}_{obs} = \tau$ $\ldots$ where $\begin{cases} \tau = \dfrac{V}{v}, \\ \bar{t}_{obs} = \dfrac{V_{active}}{v}. \end{cases}$

(b) The semilog plot is a convenient tool for evaluating the flow parameters of a mixed flow compartment. Just draw the straight line tracer response curve on this plot, find the slope and intercept, and this gives the quantities A, B, and C, as shown in Fig. 5.4.

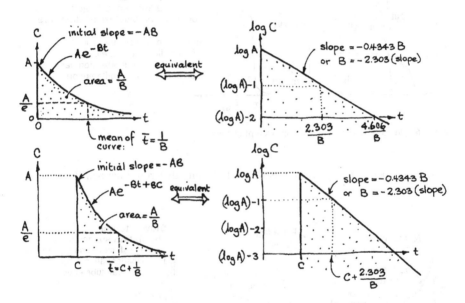

Fig. 5.4 The semi log plot

(c) Remember, we go from C to E_t as shown in Fig. 5.5.

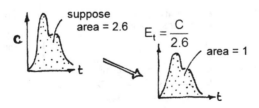

Fig. 5.5 By making A = 1 changes the C to E

(d) This type of model is useful for diagnostic purposes, to pinpoint faulty flow, and to suggest causes. For example, if you expect *plug flow* and you know $\tau = V/v$, here is what you could find (Fig. 5.6).

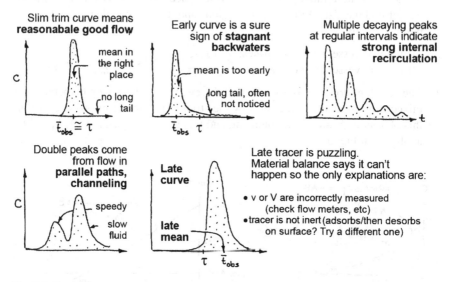

Fig. 5.6 Do these curves seem to approach plug flow?

If you expect *mixed flow* here is the sort of thing that could happen (Fig. 5.7).

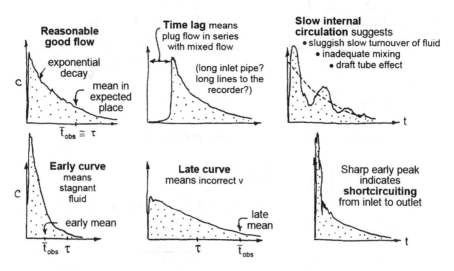

Fig. 5.7 Do these curves seem to approach mixed flow?

Example 1. *Finding the RTD by Experiment.*
The concentration readings in Table E1 represent a continuous response to a pulse input into a chemical reactor.

(a) Calculate the **E** function time of fluid in the vessel $\bar{t}$
(b) Tabulate and plot the exit age distribution **E**.

Table E1

Time t (min)	Tracer output concentration, C_{pulse} (g/L fluid)
0	0
5	3
10	5
15	5
20	4
25	2
30	1
35	0

Solution.

(a) *The mean residence time*, from Chap. 2, is

$$\bar{t} = \frac{\sum t_i C_i \Delta t_i}{\sum C_i \Delta t_i} \underline{\Delta t = \text{constant}} \frac{\sum t_i C_i}{\sum C_i}$$
$$= \frac{5 \times 3 + 10 \times 5 + 15 \times 5 + 20 \times 4 + 25 \times 2 + 30 \times 1}{3 + 5 + 5 + 4 + 2 + 1} = 15 \text{ min}.$$

Solution.

(a) *Calculate* **E**, The area under the concentration–time curve,

$$\text{Area} = \sum C\Delta t = (3 + 5 + 5 + 4 + 2 + 1)5 = 100 \text{ g} \cdot \text{min}/\text{L}$$

gives the total amount of tracer introduced. To find **E**, the area under this curve must be unity; hence, the concentration readings must each be divided by the total area, giving

$$\mathbf{E} = \frac{C}{\text{area}}, \text{ min}^{-1}. \leftarrow (a)$$

(b) *Tabulate and plot* (Fig. 5.8)

t (min)	0	5	10	15	20	25	30
$\mathbf{E} = \frac{C}{\text{area}}$ (min^{-1})	0	0.03	0.05	0.05	0.04	0.02	0.01

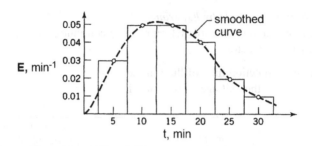

Fig. 5.8 Plot of a table of a fitled curve

Example 2. *Behavior of Liquid in a G/L Contactor.*
A large tank (860 L) is used as a G/L contactor. Gas bubbles up through the vessel
and out the top, liquid flows in at one port and out the other at 5 L/s. To get an idea
of the flow pattern of liquid in the tank, a pulse of tracer is injected at the liquid inlet
and is measured at the outlet, as shown in Fig. 5.9.

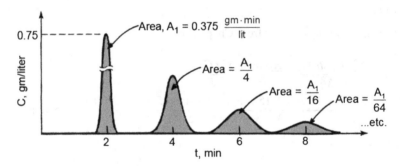

Fig. 5.9 Fraction of G and L in a contactor

(a) Find the liquid fraction of the vessel.
(b) Qualitatively, what do you think is happening in the vessel?

Solution.

(a) *Liquid fraction.* The area under the tracer curve

$$\text{Area} = \sum A_i = A_1 \left(1 + \frac{1}{4} + \frac{1}{16} + \cdots \right) = 0.375 \left(\frac{4}{3} \right) = 0.5 \frac{\text{g min}}{\text{L}}.$$

The mean residence time for the liquid from (2.5), is

$$\bar{t}_1 = \frac{\sum t_i C_i}{\sum C_i} = \frac{1}{0.5} \left[2A_1 + 4 \left(\frac{A_1}{4} \right) + 6 \left(\frac{A_1}{16} \right) + \cdots \right] = 2.67 \ \text{min}.$$

Thus, the volume of liquid from (3.2) or (5.6) is

$$V_1 = \bar{t}_1 v_1 = 2.67(5 \times 60) = 800 \text{ L.}$$

So, the fraction of phases is

$$\left. \begin{array}{l} \text{Fraction of liquid} = \frac{800}{860} = 93\% \\ \text{Fraction of gas} = 7\% \end{array} \right\} \leftarrow \text{(a)}.$$

(b) The vessel shows a strong recirculation of liquid, probably induced by the rising bubbles, see Fig. 5.10.

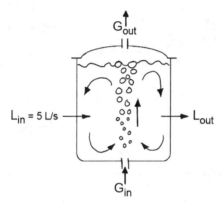

Fig. 5.10 Guessed behavior of G and L in the contractor

Problems

A pulse of NaCl solution is introduced as tracer into the fluid entering a vessel ($V = 1 \text{ m}^3$, $v = 1 \text{ m}^3/\text{min}$) and the concentration of tracer is measured in the fluid leaving the vessel. Develop a flow model to represent the vessel from the tracer output data, as shown in Fig. 5.11, (problems 5.1 to 5.8).

A step input tracer test (switching from tap water to salt water, and measuring the conductivity of fluid leaving the vessel) is used to explore the flow pattern of fluid through the vessel ($V = 1 \text{ m}^3$, $v = 1 \text{ m}^3/\text{min}$). Devise a flow model to represent the vessel from Fig. 5.12, (problems 5.9 to 5.16).

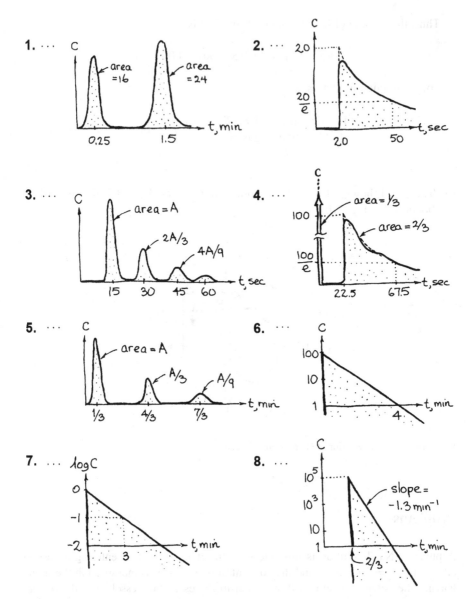

Fig. 5.11 Output for pulse tracer inputs, $V = 1\,m^3$, $v = 1\,m^3/min$

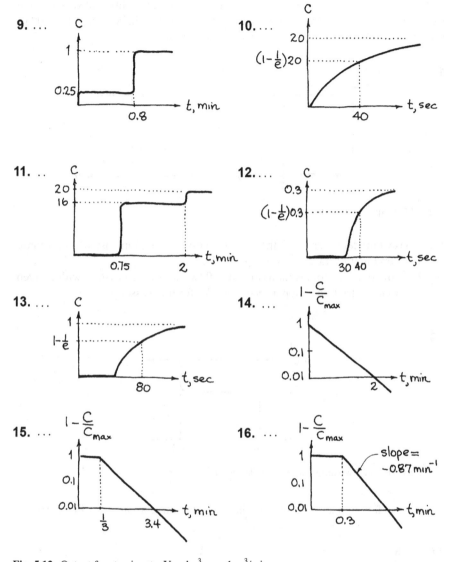

Fig. 5.12 Output for step inputs, $V = 1 m^3$, $v = 1 m^3/min$

Find a flow model which will give the normalized response curve of Fig. 5.13.

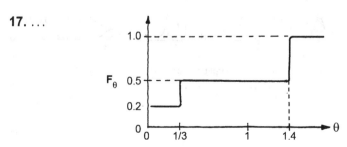

Fig. 5.13 Output to step input

Figure 5.14a and 14b show the results of tracer tests on two different vessels using step inputs of tracer (switching from salt water to tap water). In both cases $v = 100$ L/min and $V = 80$ L. Devise flow models to represent these.

18. ...

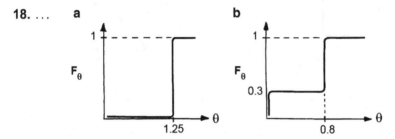

Fig. 5.14 Outputs to two step inputs

(a) Devise a model to represent the flow in a vessel whose dimensionless response to a step input of tracer is given in Fig. 5.15.
(b) To show that sometimes more than one flow model is consistent with a given tracer curve, try to develop a second model for this vessel.

19. ...

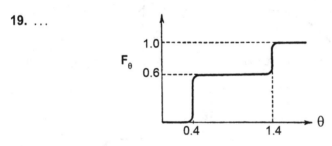

Fig. 5.15 Develop two different flow models to fit this **F** curves

Develop flow models to represent the two sets of tracer curves of Fig. 5.16.

20. ... **a**

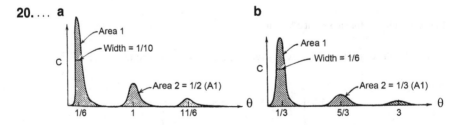

Fig. 5.16 Develop flow models for these two output curves

Chapter 6
The Dispersion Model

Models are useful for representing flow in real vessels, for scale up, and for diagnosing poor flow. We have different kinds of models depending on whether flow is close to plug, or mixed, or intermediate (Fig. 6.1).

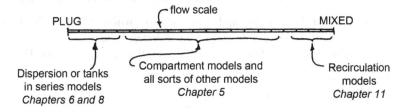

Fig. 6.1 Flow scale shows what models to use

Chapters 6–8 deal primarily with small deviations from plug flow. There are two models for this: the *dispersion model* and the *tanks in series* model. Use the model which is comfortable for you. They are roughly equivalent. These models apply to turbulent flow in pipes, laminar flow in long tubes, packed beds, shaft kilns, and long channels.

For laminar flow in short tubes or laminar flow of viscous materials, these models may not apply, and it may be that the parabolic velocity profile is the main cause of deviation from plug flow. We treat this situation called the *pure convection model* in Chap. 9.

If you are unsure of which model to use, go to Fig. 9.2. It will tell you which model should represent your set up.

6.1 Longitudinal Dispersion

To characterize the longitudinal spread of a pulse of flowing tracer, we assume a diffusion-like process superimposed on plug flow. We call this dispersion or longitudinal dispersion, with coefficient $\mathbf{D}$ (m^2/s). Take care to distinguish it from molecular diffusion, with its coefficient $\mathfrak{D}$ (m^2/s) (Fig. 6.2).

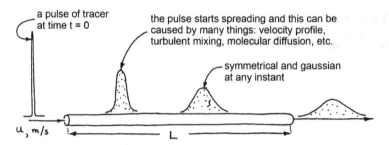

Fig. 6.2 For behavior close to plug flow

Researchers often confuse $\mathbf{D}$ with $\mathfrak{D}$, as well as their corresponding dimensionless groups. These groups are as follows:

$\mathbf{D}/uL$, called the dispersion number, characterizes the spreading rate of flowing material in the vessel caused by *all acting factors* – whether it be turbulent mixing, or by laminar velocity profile, molecular diffusion, etc.

$\mathfrak{D}/uL = \mathrm{Bo} = 1/\mathrm{Pe}$ represents the spreading caused by *molecular diffusion alone*.

$\mathfrak{D}/uL$ is called the Bodenstein number by chemists, while $uL/\mathfrak{D}$ is called Peclet number by engineers. Neither is the dispersion number, $\mathbf{D}/uL$.

The dispersion coefficient $\mathbf{D}$ represents this overall spreading process. Thus

- Large $\mathbf{D}$ means rapid spreading of the tracer curve
- Small $\mathbf{D}$ means slow spreading
- $\mathbf{D} = 0$ means no spreading; hence, plug flow.

We evaluate $\mathbf{D}$ or $\mathbf{D}/uL$ by recording the shape of the tracer curve as it passes the exit of the vessel. In particular, we measure

$\bar{t}$ = mean time of passage, or when the curve passes by
σ^2 = variance, or [how long]^2 it takes for the curve to pass by the measuring point

These measures, $\bar{t}$ and σ^2, are directly linked by theory to $\mathbf{D}$ and $\mathbf{D}/uL$. On solving the mathematics, we find a simple solution for slow spreading of tracer, but a more complex solution for rapid spreading. Let us look at these solutions in turn.

6.2   Small Deviation from Plug Flow, $\mathbf{D}/uL < 0.01$, $\sigma^2 < 0.02$, or $\sigma_t^2 < 0.02t^{-2}$

Here the tracer curve is narrow, so it passes the measuring point quickly compared to $\bar{t}$. As a result, we assume that the measured curve *doesn't change shape* as it is being measured; thus, it is symmetrical and Gaussian. It is most conveniently represented by the $\mathbf{E}_\theta$ vs. θ coordinates (see Chap. 3) with a spread which depends on $\mathbf{D}/uL$. Its characteristics, properties, and equation are as follows (Fig. 6.3).

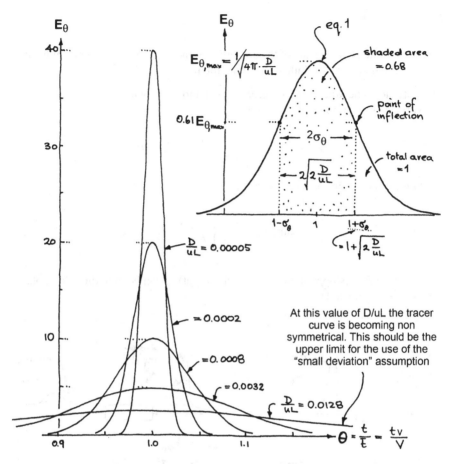

Fig. 6.3 Symmetrical response curve for very small deviation from plug flow, $\mathbf{D}/uL < 0.01$. This curve does not change shape as it is being measured

$$\mathbf{E}_\theta = \bar{t} \cdot \mathbf{E} = \frac{1}{\sqrt{4\pi(\mathbf{D}/uL)}} \exp\left[\frac{(1-\theta^2)}{4(\mathbf{D}/uL)}\right], \tag{6.1}$$

$$\mathbf{E} = \sqrt{\frac{u^3}{4\pi\mathbf{DL}}} \exp\left[-\frac{(L-ut)^2}{4(\mathbf{DL}/u)}\right],$$

Mean of the **E** curve: $\bar{t}_\mathbf{E} = \dfrac{V}{v} = \dfrac{L}{u}$ or $\bar{\theta}_\mathbf{E} = 1,$

Variance of the **E** curve: $\sigma_\theta^2 = \dfrac{\sigma_t^2}{\bar{t}^2} = 2\left(\dfrac{\mathbf{DL}}{u^3}\right)$ or $\sigma_t^2 = 2\left(\dfrac{\mathbf{DL}}{u^3}\right).$

6.2.1   Comments

(a) To evaluate D/uL use any of the properties of the pulse-response curve, either σ_t^2, maximum height, or width at 61% maximum height.

(b) Note how the tracer curve spreads as it moves down the vessel. From the variance expression of (6.1) we find

$$\sigma_t^2 \propto L \quad \text{or} \quad \left(\frac{\text{width of}}{\text{tracer curve}} \right)^2 \propto L.$$

(c) For a sloppy input of tracer we have (Fig. 6.4)

$$\left. \begin{array}{c} \bar{t} \\ \sigma^2 \end{array} \right\} \leftarrow \text{same value} \rightarrow \left\{ \begin{array}{l} \overline{\Delta t} = \bar{t}_{\text{out}} - \bar{t}_{\text{in}}, \\ \Delta\sigma^2 = \sigma_{\text{out}}^2 - \sigma_{\text{in}}^2. \end{array} \right. \tag{6.2}$$

Since $\Delta\sigma^2$ (sloppy input) $= \sigma^2$ (pulse input), it can be used directly in place of σ^2 to evaluate D/uL.

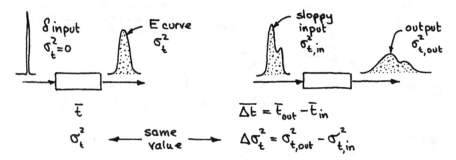

Fig. 6.4   Increase in variance is the same in both cases

(d) For a *series of vessels* the variances are additive, thus (Fig. 6.5)

$$\overline{\Delta t}_{\text{overall}} = \bar{t}_a + \bar{t}_b + \cdots = \frac{V_a}{v} + \frac{V_b}{v} + \cdots = \left(\frac{L}{u}\right)_a + \left(\frac{L}{u}\right)_b + \cdots, \tag{6.3a}$$

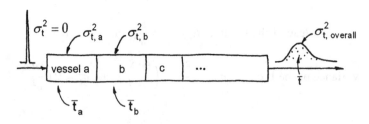

Fig. 6.5   Output for a series of vessels

$$\Delta\sigma^2_{t,\text{overall}} = \sigma^2_{t,a} + \sigma^2_{t,b} + \cdots = 2\left(\frac{\mathbf{D}L}{u^3}\right)_a + 2\left(\frac{\mathbf{D}L}{u^3}\right)_b + \cdots. \qquad (6.3b)$$

The additivity of times is expected, but the additivity of variance is not generally expected. This is a useful property since it allows us to subtract from the measured curve distortions caused by input lines, long measuring leads, etc.

6.3 Large Deviation from Plug Flow, D/uL > 0.01

Here the pulse response is broad, and it passes the measurement point slowly enough that it changes shape – it spreads – as it is being measured. This gives a nonsymmetrical **E** curve.

An additional complication enters the picture for large **D**/uL: What happens right at the entrance and exit of the vessel strongly affects the shape of the tracer curve as well as the relationship between the parameters of the curve and **D**/uL.

Let us consider two types of boundary conditions: either the flow is undisturbed as it passes the boundary (we call this the *open* b.c.), or there is plug flow outside the vessel up to the boundary (we call this the *closed* b.c.). This leads to four combinations of boundary conditions, each with its response curve (Fig. 6.6).

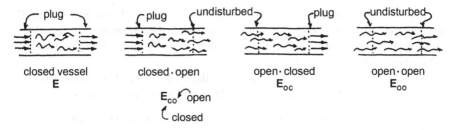

Fig. 6.6 Boundary conditions to the vessel: $\mathbf{E}$, $\mathbf{E}_{co}$, $\mathbf{E}_{oc}$, $\mathbf{E}_{oo}$

- If the fluid enters and leaves the vessel in small pipes in turbulent flow, then you have a closed b.c. (Fig. 6.7).

Fig. 6.7 Vessel with closed b.c.

- Measuring the dispersion in a section of a larger vessel or in a section of long
  pipe represents an open b.c. (Fig. 6.8).

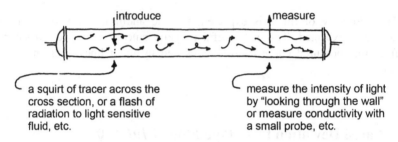

a squirt of tracer across the
cross section, or a flash of
radiation to light sensitive
fluid, etc.

measure the intensity of light
by "looking through the wall"
or measure conductivity with
a small probe, etc.

Fig. 6.8   Vessel with open b.c.

- Taking a mixing cup measurement at the exit represents a closed b.c. (Fig. 6.9).

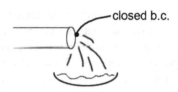

Fig. 6.9   Vessel with closed exit

Now only one boundary condition gives a tracer curve which is identical to the **E**
function, and that is the closed b.c. For all other boundary conditions, you do not get
a proper RTD, and $\bar{t} > V/v$. To distinguish between these curves we use indicating
subscripts, thus $\mathbf{E}, \mathbf{E}_{co}, \mathbf{E}_{oc}, \mathbf{E}_{oo}$.

In all cases we can evaluate $\mathbf{D}/uL$ from the parameters of the tracer curves;
however, each set of boundary conditions leads to its own mathematics.

1. *Closed vessel.* Here, an analytic expression for the **E** curve is not available.
   However, we can construct the curve by numerical methods, and evaluate its
   mean and variance exactly. Thus (Fig. 6.10) (van der Laan 1958)

$$\bar{t}_E = \bar{t} = \frac{V}{v} \text{ or } \bar{\theta}_E = \frac{\bar{t}_E}{t} = \frac{\bar{t}_E v}{V} = 1$$

$$\sigma_\theta^2 = \frac{\sigma_t^2}{\bar{t}^2} = 2\left(\frac{\mathbf{D}}{uL}\right) - 2\left(\frac{\mathbf{D}}{uL}\right)^2 \left[1 - e^{-uL/\mathbf{D}}\right]$$

(6.4)

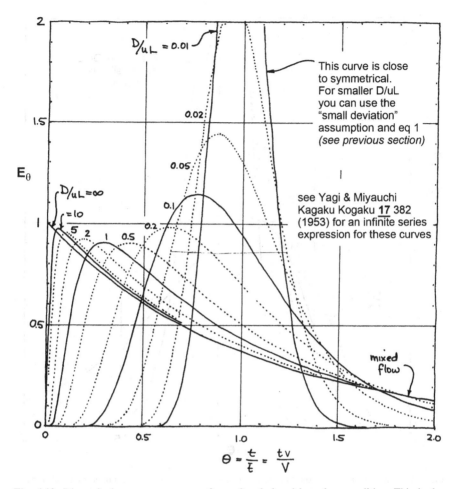

The curve labels and annotations in the figure:

$D/uL = 0.01$

This curve is close to symmetrical. For smaller D/uL you can use the "small deviation" assumption and eq 1 (see previous section)

0.02

0.05

E_θ

$D/uL = \infty$

$= 10$

$5 \ 2$

$1 \ 0.5$

0.1

0.2

see Yagi & Miyauchi Kagaku Kogaku **17** 382 (1953) for an infinite series expression for these curves

mixed flow

$$\theta = \frac{t}{\bar{t}} = \frac{t\,v}{V}$$

Fig. 6.10 Dimensionless response curves for a closed-closed boundary condition. This is the proper E_θ function

2. *Open closed and closed open vessels*. Again, we cannot derive the equation for, but we can construct the tracer curves and evaluate their mean and variance in terms of D/uL. Thus (Fig. 6.11) (van der Laan 1958)

$$\bar{\theta}_{Eoc} = \bar{\theta}_{Eco} = 1 + \frac{D}{uL} \quad \text{or} \quad \bar{t}_{Eoc} = \bar{t}_{Eco} = \frac{V}{v}\left(1 + \frac{D}{uL}\right),$$

$$\sigma^2_{\theta oc} = \sigma^2_{\theta co} = \frac{\sigma^2_{t,oc}}{\bar{t}^2} = \frac{\sigma^2_{t,oc}}{(V/v)^2} = 2\left(\frac{D}{uL}\right) + 3\left(\frac{D}{uL}\right)^2. \tag{6.5}$$

3. *Open vessel*. This represents a convenient and commonly used experimental device, a section of long pipe. It also happens to be the only physical situation (besides small D/uL) where the analytical expression for the E curve is not too complex. We thus find the following response curve and equations (Fig. 6.12) (Levenspiel and Smith 1957) and (6.6).

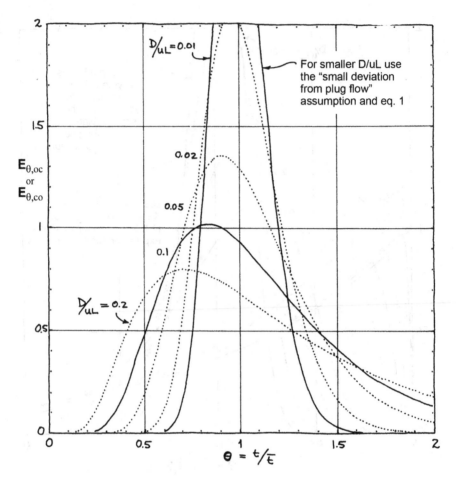

Fig. 6.11 Dimensionless response curve for open–closed or closed–open boundary conditions

$$\mathbf{E}_{\theta,\,oo} = \frac{1}{\sqrt{4\pi(\mathbf{D}/uL)\theta}}\exp\left[-\frac{(1-\theta)^2}{4\theta(\mathbf{D}/uL)}\right]$$

$$\mathbf{E}_{oo} = \frac{u}{\sqrt{4\pi\mathbf{D}t}}\exp\left[-\frac{(L-ut)^2}{4\mathbf{D}t}\right]$$

$$\bar{\theta}_{\mathbf{E}_{oo}} = \frac{\bar{t}_{\mathbf{E}_{oo}}}{\bar{t}} = 1+2\left(\frac{\mathbf{D}}{uL}\right)\ \text{or}\ \bar{t}_{\mathbf{E}_{oo}} = \frac{V}{v}\left(1+2\frac{\mathbf{D}}{uL}\right)$$

$$\sigma^2_{\theta,oo} = \frac{\sigma^2_{t,oo}}{\bar{t}^2} = 2\frac{\mathbf{D}}{uL} + 8\left(\frac{\mathbf{D}}{uL}\right)^2$$

(6.6)

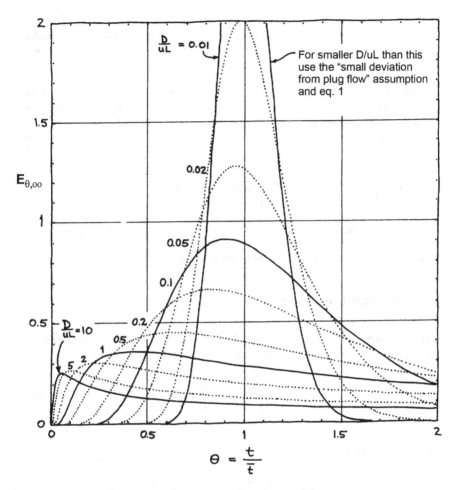

Fig. 6.12 Dimensionless response for open–open boundary conditions

6.3.1   Comments

(a) Note that the areas under the curves of these three charts are unity. Their means are not all unity.

(b) For small **D**/*uL*, the curves for the different boundary conditions all approach the "small deviation" curve of (6.1). At larger **D**/*uL* the curves differ more and more from each other.

(c) To evaluate **D**/*uL*, match either the measured tracer curve or σ^2 to theory. Matching σ^2 is simplest (though not necessarily best) so it is often used. But be sure to use the right boundary conditions.

(d) If the flow deviates greatly from plug (**D**/*uL* large), chances are that the real vessel does not meet the assumption of the model (a lot of independent random fluctuations). Here it becomes questionable whether the model should even be used. I hesitate when **D**/*uL* > 1.

(e) You must always ask whether the model should be used. You can always match σ^2 values, but if the shape looks wrong do not use this model, use some other model. For example do not use the dispersion model for the curves shown in Fig. 6.13.

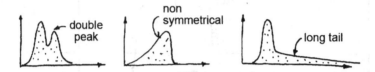

Fig. 6.13   Do not use the dispresion model for these curves

6.4   Step Input of Tracer

Here the output **F** curve is S-shaped, and it is obtained by integrating the corresponding **E** curve. Thus, at any time t or θ (Fig. 6.14)

$$\mathbf{F} = \int_0^\theta \mathbf{E}_\theta \, d\theta = \int_0^t \mathbf{E}_t \, dt. \tag{6.7}$$

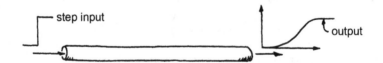

Fig. 6.14   The step-response experiment

The shape of the **F** curve depends on **D**/*uL* and the boundary conditions for the vessel. Analytical expressions are not available for any of the **F** curves; however, their graphs can be constructed, and these are shown below:

1. *Small deviation from plug flow, small dispersion,* **D**/*uL* < 0.01. From (6.1) and (6.7) we find Fig. 6.15.

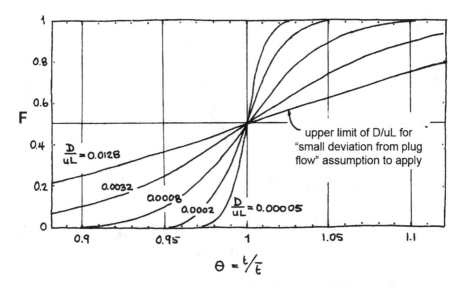

Fig. 6.15 Symmetrical step response curve for small deviation from plug flow, $D/uL < 0.01$

2. *Closed-closed vessel with large dispersion,* $D/uL > 0.01$. This is the "proper" **F** curve, the one which gives the "proper" **E** curve on differentiating (Fig. 6.16).

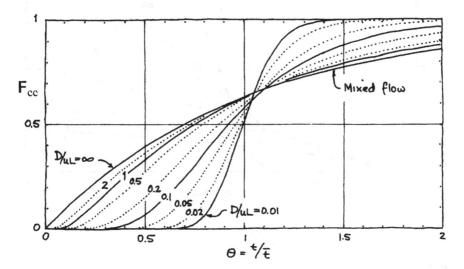

Fig. 6.16 Closed–closed step response curves

3. *Closed·open and open·closed vessels with large dispersion,* $D/uL > 0.01$. The curves are identical for these two cases (Fig. 6.17).

4. *Open vessel with large dispersion,* $D/uL > 0.01$. Integration of (6.6) gives the following F_{oo} curve (Fig. 6.18).

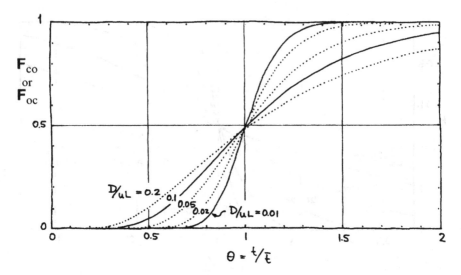

Fig. 6.17 Open–closed step response curves

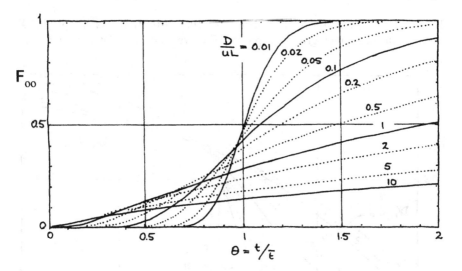

Fig. 6.18 Open–open step response curves

6.4.1   Comments

(a) To evaluate D/uL from the **F** curve either

- Match the experimental **F** curve to one of the family of curves displayed in the above figures.

- From the data calculate $\bar{t}$ and σ^2 for the corresponding pulse response curve, and use them to evaluate D/uL. See Chap. 2 for the calculation procedure.
- For small enough deviation from plug flow ($D/uL < 0.01$), plot the **F** curve directly on probability paper (see Fig 2.5 and 2.6).

(b) One direct commercial application is to find the zone of intermixing – the contaminated width – between two fluids of somewhat similar properties flowing one after the other in a long pipeline. See Chap. 7.

   To summarize: in designing a pipeline for minimum contamination use turbulent flow throughout and have no side loops or holding tanks anywhere along the line.

(c) Should you use a pulse or step injection experiment? Sometimes one type of experiment is naturally more convenient for one of many reasons. In such a situation this question does not arise. But when you do have a choice, then the pulse experiment is preferred because it gives a more "honest" result. The reason is that the **F** curve integrates effects, and gives a smooth good-looking curve which could well hide real effects. For example, the sketches below show the corresponding **E** and **F** curves for a given vessel (Fig. 6.19).

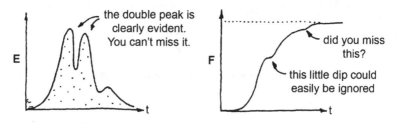

Fig. 6.19 Equivalent **E** and **F** curves

6.5 Experimental Values for Axial Dispersion

The vessel dispersion number:

$$\frac{D}{uL} = \begin{pmatrix} \text{intensity} \\ \text{of dispersion} \end{pmatrix}\begin{pmatrix} \text{geometric} \\ \text{factor} \end{pmatrix} = \frac{D}{ud} \cdot \frac{d}{L}$$

characteristic length: d_t or d_p (6.8)

this characterizes the vessel as a chemical reactor

$$\frac{D}{ud} = f\begin{pmatrix} \text{fluid} & \text{flow} \\ \text{properties'} & \text{dynamics} \end{pmatrix}$$

Schmidt no.   Reynolds no.

Theory and experiment give D/ud for various situations. We summarize these findings in Figs. 6.20–6.24.

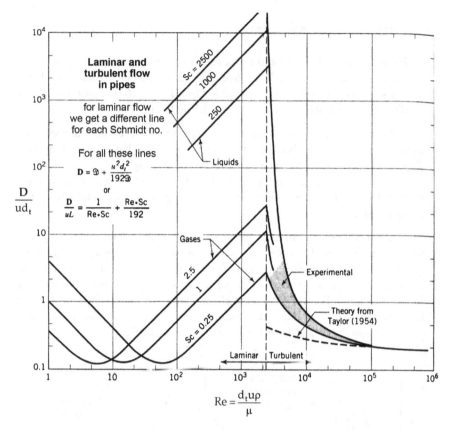

Fig. 6.20 Correlation for the dispersion of fluids flowing in pipes, adapted from Levenspiel (1958)

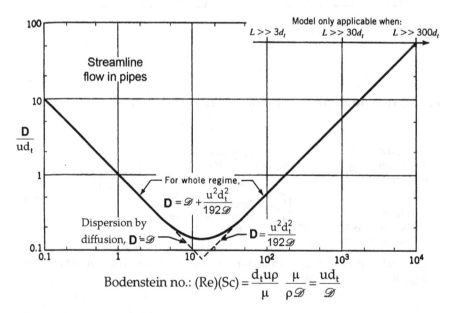

Fig. 6.21 Correlation for dispersion for streamline flow in pipes; prepared from Taylor (1953, 1954) and Aris (1956)

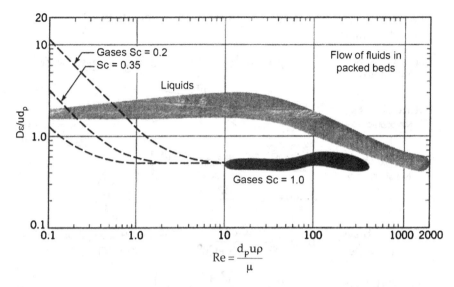

Fig. 6.22 Experimental findings on dispersion of fluids flowing with mean axial velocity u in packed beds; prepared in part from Bischoff (1961). The mean axial movement of tracer down the packed bed is $u = u_o \varepsilon$ where ε = bed voidage

6.5.1   Extensions and Comments

(a) As a rule of thumb for experimentation, we get a Gaussian RTD curve if

$$\frac{D}{uL} < 0.01 \quad \text{and} \quad V_{\substack{\text{injected} \\ \text{sample}}} < 0.01 V_{\text{vessel}}.$$

(b) In laminar flow of Newtonians (Fig. 6.25):

- If Bo < 3 use $\mathbf{D} = \mathfrak{D}$. In this regime axial dispersion is caused by molecular diffusion alone
- If Bo > 50 use $\mathbf{D} = u^2 d_t^2 / 192 \mathfrak{D}$. In this regime axial dispersion is a result of the interaction of velocity variations with lateral molecular diffusion.
- If 3 < Bo < 50 use $\mathbf{D} = \mathfrak{D} + u^2 d_t^2 / 192 \mathfrak{D}$. This is the intermediate regime where all mechanisms intrude.

   The charts verify these conclusions. These relationships were derived theoretically by Taylor (1953, 1954) and Aris (1956).

(c) Finding $\mathbf{D}$ of fluids by experiment. Measuring the axial dispersion in laminar flow is a neat, quick, simple, and accurate way for finding the molecular diffusion coefficient of fluids. However, in solving the quadratic expression you will come up with two $\mathbf{D}$ values. In other words, two $\mathbf{D}$ values will give the same extent of axial dispersion (Fig. 6.25).

   If you know which arm of the curve you are on (a rough estimate of the Bodenstein number may help), you can find the correct value for $\mathbf{D}$. If not, then you cannot tell and you need at least two runs at different velocities or in different size pipes.

Laminar flow in all shapes of channels and conduits give closely similar results

Circular tubes,
Newtonians
$$D = \mathcal{D} + \frac{u^2 d_t^2}{192\,\mathcal{D}}$$
··· Aris, Proc. Roy. Soc. 235A 67 (1956)

Parallel slots,
Newtonians
$$D = \mathcal{D} + \frac{u^2 d^2}{210\,\mathcal{D}}$$
Aris,
Proc. Roy. Soc.
252A 538 (1959)

Elliptical channels,
Newtonians
$$D = \mathcal{D} + \frac{24 - 24e^2 + 5e^4}{24 - 12e^2} \cdot \frac{u^2 a^2}{192\,\mathcal{D}}$$
where $e^2 = \dfrac{a^2 - b^2}{a^2}$

Circular tubes,
power law fluids
$$D = \mathcal{D} + \frac{u^2 d_t^2}{8(n+3)(n+5)\,\mathcal{D}}$$
··· Fan and Hwang, Proc. Roy. Soc
283A 576 (1965)

Parallel slots,
power law fluids
$$D = \mathcal{D} + \frac{u^2 d^2}{6(n+4)(2n+5)\,\mathcal{D}}$$

Circular tubes,
Bingham plastics
see eq. 17 in the reference
(rather long)

··· Fan and Wong
Proc. Roy. Soc
292A 203 (1966)

(e) For time variable flow, say in an estuary or an artery, use a time average **D**, see Bischoff, CES 19 989 (1964)

(f) Newtonians in coiled tubes. In certain flow regimes dispersion is higher than in straight pipe, in other regimes it is lower.

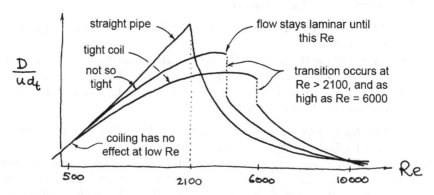

Fig. 6.23 Coiled tubes give lower and higher dispresion than do straight pipes

Adsorbing porous-walled tubes, and a tube model for packed beds of porous adsorbing particles

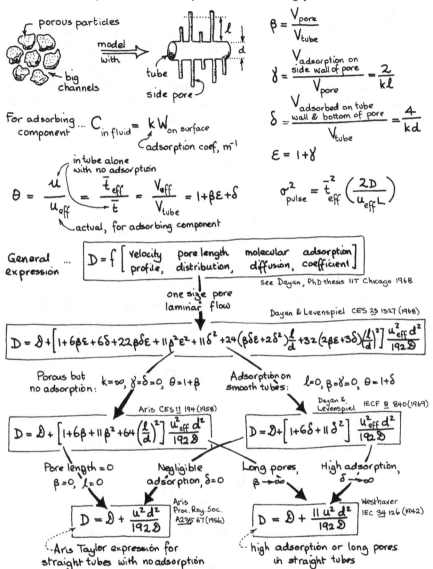

porous particles

big channels

model with

tube

side pore

$\beta = \dfrac{V_{pore}}{V_{tube}}$

$\gamma = \dfrac{V_{adsorption\ on\ side\ wall\ of\ pore}}{V_{pore}} = \dfrac{2}{k\ell}$

For adsorbing component ... $C_{in\ fluid} = k\,W_{on\ surface}$

adsorption coef, m^{-1}

$\delta = \dfrac{V_{adsorbed\ on\ tube\ wall\ \&\ bottom\ of\ pore}}{V_{tube}} = \dfrac{4}{kd}$

$\varepsilon = 1 + \gamma$

in tube alone with no adsorption

$\theta = \dfrac{u}{u_{eff}} \left(= \dfrac{\bar{t}_{eff}}{\bar{t}} = \dfrac{V_{eff}}{V_{tube}} \right) = 1 + \beta\varepsilon + \delta$

actual, for adsorbing component

$\sigma^2_{pulse} = \bar{t}^2_{eff} \left(\dfrac{2D}{u_{eff}\,L} \right)$

General expression ...

$$D = f \left[\begin{array}{l} velocity\ \ pore\ length\ \ \ molecular\ \ adsorption \\ profile, \ \ distribution, \ \ diffusion, \ \ coefficient \end{array} \right]$$

See Dayan, PhD thesis IIT Chicago 1968

one size pore laminar flow

Dayan & Levenspiel CES 23 1327 (1968)

$$D = \mathcal{D} + \left[1 + 6\beta\varepsilon + 6\delta + 22\beta\delta\varepsilon + 11\beta^2\varepsilon^2 + 11\delta^2 + 24\left(\beta\delta\varepsilon + 2\delta^2\right)\frac{\ell}{d} + 32\left(2\beta\varepsilon + 3\delta\right)\left(\frac{\ell}{d}\right)^2 \right] \dfrac{u^2_{eff}\,d^2}{192\,\mathcal{D}}$$

Porous but no adsorption: $k=\infty,\ \gamma=\delta=0,\ \theta=1+\beta$

Adsorption on smooth tubes: $\ell=0,\ \beta=\gamma=0,\ \theta=1+\delta$

Dayan & Levenspiel IECF 8 840 (1969)

Aris CES 11 194 (1958)

$$D = \mathcal{D} + \left[1 + 6\beta + 11\beta^2 + 64\left(\frac{\ell}{d}\right)^2 \right] \dfrac{u^2_{eff}\,d^2}{192\,\mathcal{D}}$$

$$D = \mathcal{D} + \left[1 + 6\delta + 11\delta^2 \right] \dfrac{u^2_{eff}\,d^2}{192\,\mathcal{D}}$$

Pore length = 0 $\beta=0,\ \ell=0$

Negligible adsorption, $\delta=0$

Long pores, $\beta \to \infty$

High adsorption, $\delta \to \infty$

Aris Proc. Roy. Soc. A235 67 (1956)

$$D = \mathcal{D} + \dfrac{u^2 d^2}{192\,\mathcal{D}}$$

Westhaver IEC 34 126 (1942)

$$D = \mathcal{D} + \dfrac{11\,u^2 d^2}{192\,\mathcal{D}}$$

Aris Taylor expression for straight tubes with no adsorption

high adsorption or long pores in straight tubes

Fig. 6.24 Dispresion in packed beds of porous particles

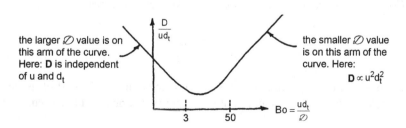

$\dfrac{D}{ud_t}$

the larger $\mathcal{D}$ value is on this arm of the curve. Here: **D** is independent of u and d_t

the smaller $\mathcal{D}$ value is on this arm of the curve. Here: $D \propto u^2 d_t^2$

3   50

$Bo = \dfrac{ud_t}{\mathcal{D}}$

Fig. 6.25 From von Rosenberg (1956)

Example 1. **D**/*uL from a C Curve.*

On the assumption that the closed vessel of Example 3.1 is well represented by the dispersion model, calculate the vessel dispersion number **D**/*uL* (see Chap. 3).

Solution. Since the C curve for this vessel is broad and unsymmetrical, let us guess that dispersion is too large to allow use of the simplification leading to Fig. 6.1. We thus start with the variance matching procedure. The variance of a continuous distribution measured at a finite number of equidistant locations is given by (2.6) as

$$\sigma^2 = \frac{\sum t_i^2 C_i}{\sum C_i} - \bar{t}^2$$

Using the original tracer concentration-time data given in Example 3.1, we find

$$\sum C_i = 3 + 5 + 5 + 4 + 2 + 1 = 20,$$

$$\sum t_i C_i = (5 \times 3) + (10 \times 5) + \cdots + (30 \times 1) = 300 \text{ min},$$

$$\sum t_i^2 C_i = (25 \times 3) + (100 \times 5) + \cdots + (900 \times 1) = 5,450 \text{ min}^2.$$

Therefore

$$\sigma^2 = \frac{5,450}{20} - \left(\frac{300}{20}\right)^2 = 47.5 \text{ min}^2.$$

From Eq 2.5 we also find that $\bar{t} = \sum t_i C_i / \sum C_i = 300/20 = 15$ min.
So $\sigma_\theta^2 = \sigma^2/\bar{t}^2 = 47.5/(15)^2 = 0.211.$
Now for a closed vessel, (6.4) relates the variance to **D**/*uL*. Thus

$$\sigma_\theta^2 = 0.211 = 2\frac{\mathbf{D}}{uL} - 2\left(\frac{\mathbf{D}}{uL}\right)^2 \left(1 - e^{-uL/D}\right).$$

Ignoring the second term on the right, we have as a first approximation

$$\frac{\mathbf{D}}{uL} \cong 0.106$$

Correcting for the term ignored we find by trial and error that

$$\frac{D}{uL} = 0.120$$

From after (6.3) our original guess was correct: This value of **D**/*uL* is much beyond the limit where the simple Gaussian approximation should be used.

Example 2. **D**/*uL from an F Curve.*

Von Rosenberg (1956) studied the displacement of benzene by *n*-butyrate in a 1½-in.-diameter packed column 4 ft long, (Fig. 6.26) measuring the fraction of *n*-butyrate in the exit stream by refractive index methods. When graphed, the fraction of *n*-butyrate vs. time was found to be S-shaped. This is the **F** curve, and is shown in Fig. 6.27 for the run at the lowest flow rate where $u = 2.19 \times 10^{-5}$ ft/s, which is about 2 ft/day (refer to Chap. 3).

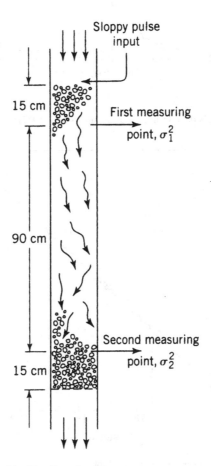

Fig. 6.26 Equipment used by Von Rosenberg

Find the vessel dispersion number for this system.

Solution. Instead of finding the *C* curve by taking the slopes of the **F** curve and then determining the spread of this curve, let us illustrate a short cut which can be used when **D**/*uL is small.

When **D**/*uL is small the *C* curve approaches the Gaussian of Fig 6.3, and the corresponding **F** curve, when plotted on probability paper, lies on a straight line. Plotting the original **F**-curve data on probability paper does actually give close to a straight line, as shown in Figs. 2.5, 2.6, or 6.28.

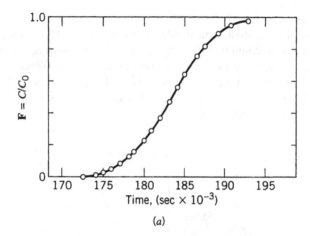

(a)

Fig. 6.27   Van Rosenbefg's data, from Levenspiel and Smith (1957)

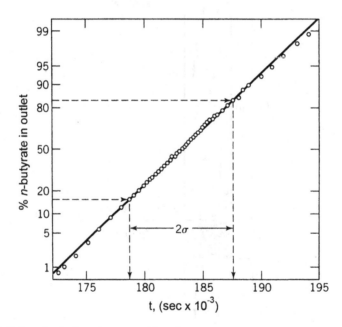

Fig. 6.28   S shaped output curve on a semi log plot

To find the variance and D/uL from a probability graph is a simple matter if we observe the following property of a Normal curve: That one standard deviation, σ, on either side of the mean of the curve includes 68% of the total area under the curve. Hence, the 16th and 84th percentile points of the **F** curve are two standard deviations apart. The 84th percentile intersects the straight line through the data at 187,750 s and the 16th percentile intersects it at 178,550 s, so the difference, 9,200 s, is taken as the value of two standard deviations. Thus, the standard deviation is

$$\sigma = 4,600\text{s}.$$

We need this standard deviation in dimensionless time units if we are to find **D**. Therefore

$$\sigma_\theta = \frac{\sigma}{\bar{t}} = (4,600\text{s}\,)\left(\frac{2.19 \times 10^{-5}\ \text{ft/s}}{4\ \text{ft}}\right) = 0.0252.$$

Hence the variance

$$\sigma_\theta^2 = (0.0252)^2 = 0.00064$$

and from (6.1)

$$\frac{\mathbf{D}}{uL} = \frac{\sigma_\theta^2}{2} = 0.00032.$$

Note that the value of **D**/uL is well below 0.01, justifying the use of the Normal approximation to the C curve and this whole procedure.

Example 3. D/uL from a One-Shot Input.
Find the vessel dispersion number in a fixed-bed reactor packed with 0.625 cm catalyst pellets. For this purpose tracer experiments are run in equipment shown in Fig. 6.26.

Catalyst is laid down in a haphazard manner above a screen to a height of 120 cm, and the fluid flows downward through this packing. A sloppy pulse of radioactive tracer is injected directly above the bed, and output signals are recorded by Geiger counters at two levels in the bed 90 cm apart.
The following data apply to a Specific experimental run.
Bed voidage = 0.4.
Superficial velocity of fluid (based on an empty tube) = 1.2 cm/s,
Variances of output signals are found to be $\sigma_1^2 = 39\ \text{s}^2$ and $\sigma_2^2 = 64\ \text{s}^2$.
Find **D**/uL.

Solution. Bischoff (1961) has shown that as long as the measurements are taken at least 2 or 3 particle diameters into the bed then the open vessel boundary conditions hold closely. This is the case here since the measurements are made 15 cm into the bed. As a result, this experiment corresponds to a one-shot input to an open vessel for which (6.2) holds. Thus

$$\Delta\sigma^2 = \sigma_2^2 - \sigma_1^2 = 64 - 39 = 25\text{s}^2$$

or in dimensionless form

$$\Delta\sigma_\theta^2 = \Delta\sigma^2\left(\frac{v}{V}\right)^2 = (25\ \text{s}^2)\left[\frac{1.2\ \text{cm/s}}{(90\ \text{cm})(0.4)}\right]^2 = \frac{1}{36}$$

from which the dispersion number is

$$\frac{\mathbf{D}}{uL} = \frac{\Delta\sigma_\theta^2}{2} = \underline{\underline{\frac{1}{72}}}$$

Problems

1. Denmark's longest and greatest river, the Gudenaa, certainly deserves study, so pulse tracer tests were run on various stretches of the river using radioactive Br-82. Find the axial dispersion coefficient in the upper stretch of the river, between Tørring and Udlum, 8.7 km apart, from the following reported measurements

Data from Danish Isotope Center, report of November 1976

t (h)	C (arbitrary)	
3.5	0	
3.75	3	
4	25	
4.25	102	
4.5	281	Pulse tracer input is at t = 0 h
4.75	535	
5	740	
5.25	780	
5.5	650	
5.75	440	
6	250	
6.25	122	
6.5	51	
6.75	20	
7	9	
7.25	3	
7.5	0	

2. RTD studies were carried out by Jagadeesh and Satyranarayana (1972) in a tubular reactor ($L = 1.21$ m, 35 mm ID). A squirt of NaCl solution (5N) was rapidly injected at the reactor entrance, and mixing cup measurements were taken at the exit. From the following results, calculate the vessel dispersion number; also the fraction volume taken up by the baffles.

t (s)	NaCl in sample	
0–20	0	
20–25	60	
25–30	210	
30–35	170	
35–40	75	($v = 1{,}300$ mL/min)
40–45	35	
45–50	10	
50–55	5	
55–70	0	

3. A pulse of radioactive Ba-140 was injected into a 10 in. pipeline (25.5 cm ID) 293 km long used for pumping petroleum products ($u = 81.7$ cm/s, Re $= 24{,}000$) from Rangely, Colorado, to Salt Lake City, Utah. Estimate the time of passage of fluid having more than $1/2C_{max}$ of tracer and compare the value you find with the reported time of passage of 895 s averaged over five runs.

    Data from Hull and Kent (1952).

4. An injected slug of tracer material flows with its carrier fluid down a long straight pipe in dispersed plug flow. At point A in the pipe the spread of tracer is 16 m. At point B 1 km downstream from A its spread is 32 m. What do you estimate its spread and its variance to be at point 2 km downstream from point A?

5. At present we are processing a liquid stream in a tubular reactor. We plan to quadruple the processing rate keeping $\bar{t}$ fixed, and for this we have two alternatives:

   - Quadruple the length of reactor leaving the pipe diameter unchanged
   - Double the pipe diameter keeping the length unchanged

   Compare the deviation from plug flow of these proposed larger units with that of the present unit. Assume that the reactors are long enough for the dispersion model to apply and that

   (a) Highly turbulent flow prevails throughout
   (b) Laminar flow prevails throughout

6. A strongly colored 1% solution of $KMnO_4$ flows slowly through a 0.504 mm ID glass tube, and at time $t = 0$ the flow is switched to pure water. The spreading front between fluids is measured 31 cm downstream by comparing the color with standardized sample tubes of diluted $KMnO_4$ solution with results shown below. From this information calculate the molecular diffusion coefficient of $KMnO_4$ in water in this concentration range. Compare your results with literature values of

$$\mathbf{D} = 4.35 \times 10^{-10} \text{ m}^2/\text{s to } 15 \times 10^{-10} \text{ m}^2/\text{s}.$$

The data for this problem comes from Taylor's classic experiment, reported in (1953).

$1 - C/C_{initial}$	t (s)
0.01	10,300
0.02	10,405
0.04	10,527
0.1	10,636
0.2	10,846
0.3	10,996
0.4	11,047
0.5	11,291
0.6	11,399
0.7	11,435
0.8	11,621
0.9	11,931

7. From the measured pulse tracer response curves, find the fraction of gas, of flowing liquid, and of stagnant deposits in the gas-liquid contactor shown in Fig. 6.29

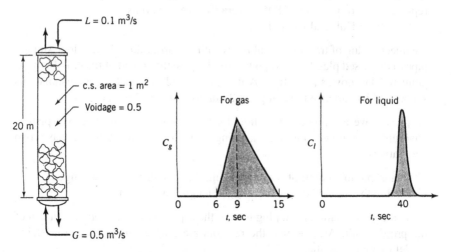

Fig. 6.29   Tracer responses for G and for L, in a G/L contactor

References

Aris, R., Proc. Roy. Soc. 235A 67 (1956). 252A 538 (1959). CES 11 194 (1958).

Bischoff, K. B., PhD Thesis, IIT Chicago (1961). CES 19 989 (1994). and O. Levenspiel, CES 17 245 (1962).

Dayan, J., PhD Thesis, IIT Chicago (1968). and O. Levenspiel, CES 23 1327 (1968). IEC/F 8 840 (1969).

Fan, L. T. and Hwang, Proc. Roy. Soc. 238A 576 (1965).

Fan, L. T., and Wong, Proc. Roy Soc., 292A 203 (1966).

Hull and Kent, IEC 44 2745 (1952).

Jagadeeth and Satyranarayana, IEC/PDD 11 520 (1972).

Van der Laan, E. T., CES 7 187 (1958).

Levenspiel, O., IEC 50 343 (1958). and W. K. Smith, CES 6 277 (1957).

Von Rosenberg, D. U., AIChE J 2 55 (1956).

Taylor, G. I. Proc. Roy. Soc. 219A 186 (1953). 225A 473 (1954).

Westhaver, IEC 126 (1972).

Yagi and Miyauchi, Kagaku Kogaku, 17 382 (1953).

Chapter 7
Intermixing of Flowing Fluids

When different fluids are pumped successively in a pipeline, switching from one to the other forms a zone of contamination W between the two, as shown in Fig. 7.1.

The width of this zone is related to the axial dispersion and the pipe dimensions which, in turn, are correlated with the Reynolds number. With the following design charts, we evaluate this width. The overall design aspects are not considered here.

Turbulent flow gives the correlation of dispersion numbers with Reynolds numbers as shown in Fig. 7.2. For Reynolds numbers above 10,000 the theoretical curve is shown and it brings together much data from thde laboratory and the field for both gases and liquids. Pipe sizes varied from 0.5 mm to 1 m in diameter and lengths as great as 650 km were used. As the Reynolds number is progressively decreased below 10,000, however, experimental studies in this range to data show an increasing scatter from this curve. This may be expected for, in the development of the theory, the laminar sublayer close to the wall is ignored and it is precisely in this region (Re < 10,000) that the sublayer becomes appreciable in thickness. This may result in the additional contribution of molecular diffusion in the sublayer as a means of axial transport.

Streamline flow is of little interest in product pipeline design. In this region of flow the zone of contamination is much greater than for turbulent flow. However, the correlation of dispersion number for streamline flow is shown in Fig. 7.3. Here molecular diffusion plays a significant role in the mixing of fluids.

For short pipes the diffusion mechanism is not applicable. Figure 7.3 should be used only when the following condition is fulfilled:

$$(Bo) = (Re)(Sc) \ll 30\frac{L}{d}. \tag{7.1}$$

The few data available check these theoretical predictions.

The *contaminated zone* is related to the system variables as shown in Fig. 7.4. The method used for preparing this figure is described in the previous chapter. The following problems illustrate the various uses of these charts.

O. Levenspiel, *Tracer Technology*, Fluid Mechanics and Its Applications 96,
DOI 10.1007/978-1-4419-8074-8_7, © Springer Science+Business Media, LLC 2012

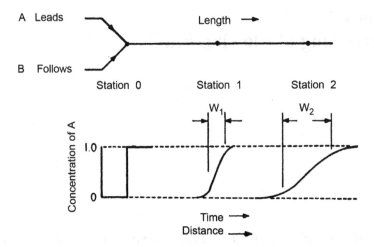

Fig. 7.1   When one fluid (**A**) is followed by another (**B**), there is a zone of mixing between the two fluids. We call this the contaminated zone

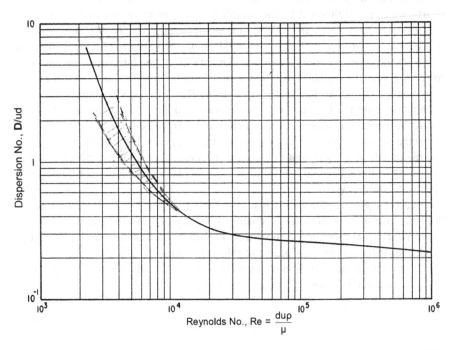

Fig. 7.2   For Reynolds numbers above 2,000 the dispersion number is a function of the Reynolds number. Here is the relation for this turbulent region of flow

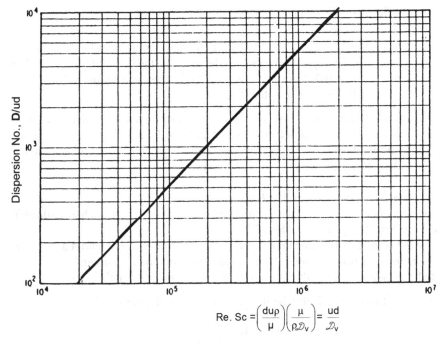

$$\text{Re. Sc} = \left(\frac{du\rho}{\mu}\right)\left(\frac{\mu}{\rho\mathcal{D}_v}\right) = \frac{ud}{\mathcal{D}_v}$$

Fig. 7.3 Here is the chart for getting the dispersion number for streamline flow. It is only applicable when (Re) (Sc) $\ll$ 30 (L/d)

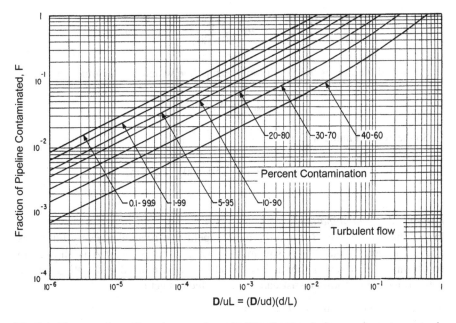

Fig. 7.4 The example problems bear out that streamline flow results in a much greater contamination zone than does turbulent flow

7.1   Conclusions

1. In general, for small D/uL ($D/uL < 0.01$) the contaminated width varies as the 0.5 power of the pipe length as may be seen by the slopes in Fig. 7.4. For example, if the length of the pipeline is made 16 times as long, the contaminated width would increase by a factor of 4.

2. Example 3 shows that laminar flow gives more than 160 times as much intermixing as does turbulent flow. So pipelines which must pump various different fluids one after the other should always be designed to operate in the turbulent flow regime, never in the laminar.

Example 1. *Contamination Width in a Pipeline.*
A variety of petroleum products are pumped one after the other from here to there in a $d = 0.5$ m ID pipeline. See Fig. 7.1.
   What is the 10–90% contaminated width as the front between fluid A and fluid B passes the measuring station 50 km downstream from the feed origin?
Both fluids have practically the same flow conditions, Re $\cong 10^5$.

Solution. Since Re $> 2{,}300$ the flow is turbulent, then from Fig. 7.2

$$\frac{D}{ud} \cong 0.26$$

from which

$$\frac{D}{uL} = \frac{D}{ud} \times \frac{d}{L} = 0.26\left(\frac{0.5}{50{,}000}\right) = 2.6 \times 10^{-6}.$$

Figure 7.4 then gives the fraction of the pipeline which is contaminated

$$F_{10-90} = 5.9 \times 10^{-3}.$$

So the contaminated width is

$$W_{10-90} = 5.9 \times 10^{-3}(50{,}000) = \underline{\underline{295\,\text{m}}}.$$

Example 2. *Contamination.*
In switching from A to B in the 0.5 m ID pipeline of Example 1, very little B is allowed to contaminate the A stream, at most 1%, but a lot of A is allowed to contaminate the B stream, about 20%. Find the contaminated width W_{1-80} at the measuring station 1 km downstream from the feed injection.

Solution. As in Example 1, $D/ud \cong 0.26$
       Then for the whole 1 km section of pipeline

$$\frac{D}{uL} = \frac{D}{ud}\left(\frac{d}{L}\right) = 0.26\left(\frac{0.5}{1{,}000}\right) = 1.3 \times 10^{-4}.$$

From Fig. 7.4 we read

$$F_{1-99} = 7.1 \times 10^{-2},$$

$$F_{20-80} = 2.8 \times 10^{-2}.$$

Since $\mathbf{D}/uL < 0.01$, the tracer curve is symmetrical, so

$$F_{1-80} = \frac{7.1 \times 10^{-2}}{2} + \frac{2.8 \times 10^{-2}}{2} = 5 \times 10^{-2}.$$

Hence, the contaminated width

$$W_{1-80} = F_{1-80}(L) = 5 \times 10^{-2}(1,000) = \underline{\underline{50\,m}}.$$

Example 3. *A Design.*
One step in the expansion of our Oregon winery involves the construction of a large tank farm to hold various types of red and white raw grape juice, all having $Sc \cong 10^3$. These are to be pumped successively to the winery and bottling plant 25 km away through a 2.5 cm ID PVC plastic pipe at $Re = 3,200$.

However, the chief wine taster suggests that in place of the 2.5 cm pipe (pipe 1) that we install a 5.0 cm pipe (pipe 2). He claims that with slower "gentler" flow the juice would be less stressed and the resulting wine would taste better. Also, there would be less mixing of the different juices, thus a smaller 1–99% contamination zone.

For the same volumetric flow rate through the pipes, determine:

(a) The contaminated widths in the two pipes on switching from one grape juice to another, W_1 and W_2.
(b) The ratio of contaminated volumes of mixture in the two pipes, V_1/V_2. This mixture represents a second class Vin Rosé.
(c) Which size pipe would you suggest we use?

Solution.

(a) For the same volumetric flow rate in the two pipes (pipe 1 = 2.5 cm ID, pipe 2 = 5.0 cm ID)

$$u_1 d_1^2 = u_2 d_2^2 = \text{constant}$$

or

$$\frac{u_1}{u_2} = \frac{d_2^2}{d_1^2} = \frac{5^2}{(2.5)^2} = 4.$$

Thus

$$\frac{Re_1}{Re_2} = \frac{(du\rho/\mu)_1}{(du\rho/\mu)_2} = \frac{d_1 u_1}{d_2 u_2} = \frac{2.5 \times 4}{5 \times 1} = 2$$

So for the 2.5 cm pipe $Re_1 = 3,200$ turbulent flow and for the 5.0 cm pipe $Re_2 = 1,600$ laminar flow.

For turbulent flow, pipe 1, d = 2.5 cm
From Fig. 7.2 for Re = 3,200

$$\left(\frac{D}{ud}\right)_1 = 2.8$$

So for the whole 25 km pipe line

$$\left(\frac{D}{uL}\right)_1 = \left(\frac{D}{ud}\right)_1 \frac{d}{L} = 2.8\left(\frac{2.5}{2,500,000}\right) = 2.8 \times 10^{-6}$$

From Fig. 7.4 the fraction of pipe line contaminated

$$(F_{1-99})_1 = 1.05 \times 10^{-2}$$

So

$$(W_{1-99})_1 = F_1(L) = 0.0106(25,000) = \underline{265 \text{ m}}$$

For laminar flow, pipe 2, d = 5.0 cm
To use Fig. 7.3 we must satisfy the condition of (7.1). Replacing values in (7.1) gives

$$10^3(1.6 \times 10^3) \overset{?}{\underset{\ll}{}} 30\left(\frac{2,500,000}{5}\right)$$

or $1.6 \times 10^6 \ll 15 \times 10^6$, so the condition satisfied. This means that we can use Fig. 7.3, which gives

$$\left(\frac{D}{ud}\right)_2 = 8 \times 10^3$$

and for the whole pipe

$$\left(\frac{D}{uL}\right)_2 = \left(\frac{D}{ud}\right)_2 \frac{d}{L} = 8 \times 10^3\left(\frac{5}{2,500,000}\right) = 0.016$$

From Fig. 7.4

$$(F_{1-99})_2 = 0.8$$

and

$$W_{1-99\,2} = F_2(L) = 0.8(25,000) = \underline{\underline{20,000 \text{ m}}}$$

(b) Find the volume ratio of wasted juices
From part (a)

$$\frac{V_1}{V_2} = \frac{W_1}{W_2}\left(\frac{(\text{c.s. area})_1}{(\text{c.s. area})_2}\right) = \frac{505}{20,000}\frac{1}{4} = 0.0063$$

So the contaminated volume in the 5 cm pipe is 160 times as large as the contaminated volume in the 2.5 cm pipe!

(c) Which pipe should we install?

The smaller diameter pipe is cheaper to install, needs a larger pump but gives a much smaller waste of juices.
Certainly we should choose the 2.5 cm pipe.

Problems

1. Kerosene and gasoline are pumped successively at 1.1 m/s through a 10 in. (25.5 cm ID) pipeline 100 km long. Calculate the 5/95–95/5% contaminated width at the exit of the pipe given that the kinematic viscosity for the 50/50% mixture is

$$u/\rho = 0.9 \times 10^{-6} \text{ m}^2/\text{s}$$

Data and problem from Sjenitzer (1958).

2. A refinery pumps products A and B successively to receiving stations 100 km away through a 10 cm ID pipeline. The average properties of A and B are $\rho = 850 \text{ kg/m}^3$, $\mu = 1.7 \times 10^{-3} \text{ kg/m} \cdot \text{s}$, $D = 10^{-9} \text{ m}^2/\text{s}$, the fluid flows at $u = 20 \text{ cm/s}$, and there are no reservoirs, holding tanks or pipe loops in the line; just a few bends.

a) Estimate the 1–90% contaminated width 100 km away.
b) If the flow is speeded to Re $= 100,000$ what would this do to the contaminated width 100 km away?

c) If the flow is slowed to Re = 1,000 what would this do to a contaminated width 100 km away? Before trying to solve this problem be sure to check pages 1, 2, and 3 of Chap. 9.

Adapted from Petroleum Refiner (1958) and Pipe Line Industry (1958).

3. 1N ammonium chloride flows at constant velocity through a straight tube 0.02371 cm ID. At time $t = 0$ we switch to 3N solution and we monitor the concentration of NH$_4$Cl 101 cm downstream with the following results

% Second fluid	5	10	40	80	97
Time, s	5,805	5,840	5,940	6,050	6,150

At the temperature of the experiment (23°C) 1–3N NH$_4$Cl have the following properties

$$\mu = 9.2 \times 10^{-4} \ kg/m \cdot s, \ \rho = 1,020 \ kg/m^3.$$

From this experiment calculate the molecular diffusion coefficient for NH$_4$Cl in water in this concentration range. How does it compare with the reported value of $\mathcal{D} = 18 \times 10^{-10} \ m^2/s$?

Data from Blackwell (1957).

4. *Diffusivity of dissolved gases in liquids.* Deaerated distilled water flows by gravity through a 1.63 mm tube wound in a 0.3 m diameter coil and immersed in a water bath kept at 17°C. A small amount of water saturated with dissolved CO$_2$ is pulsed into the tube and the concentration of CO$_2$ is monitored by a refractometer 7.5444 m downstream. At each flow rate, a bell-shaped curve is recorded. The data is as follows

Flow velocity u (mm/s)	Variance of bell-shaped curve σ_t^2 (s^2)
8.6	10,700
13.7	4,700
17.1	2,700

Calculate the diffusion coefficient of dissolved CO$_2$ in water. Any comments? Data extracted from Pratt et al. (1973).

5. A pipeline 100 km long will be constructed to transport wine from a wine-producing center to a distribution point. Red and white wine are to flow in turn through this pipeline. Naturally, in switching from one to the other, a region of *vin rosé* is formed. The quantity of *vin rosé* is to be minimized since it is not popular and does not fetch a good price on the market.

(a) How does the pipeline size at given Reynolds number affect the quantity of *vin rosé* formed in the switching operation?

(b) For fixed volumetric flow rate in the turbulent flow region, what pipe size minimizes the *vin rosé* formed during the switching operation?

(c) Assuming that the pipeline is operating at present, what flow rate should we select to minimize the formation of *vin rosé*?

6. Often a pipeline must transport more than one material. These materials are then transported successively, and switching from one to another forms a zone of contamination between the two flowing fluids. Let A refer to the leading fluid and let B refer to the following fluid in a 30-cm-i.d. pipe.
   (a) If the average Reynolds number of the flowing fluids is 10,000, find the 10–90% contaminated width 10 km downstream from the point of feed.
   (b) Find the 10–99% contaminated width at this location (10–99% contamination means allowing up to 10% of B in A but only allowing 1% of A in B).
   (c) Find the 10–99% contaminated width 160 km downstream from the point of feed.
   (d) For a given flow rate, how does contaminated width vary with length of pipe.
   For additional readings see Petroleum Refiner (1958).

7. A strongly colored 1% solution of $KMnO_4$ flows slowly through a 0.504 mm ID glass tube, and at time $t = 0$ the flow is switched to pure water. The spreading front between fluids is measured 31 cm downstream Taylor (1963), by comparing the color with standardized sample tubes of diluted $KMnO_4$ solution with results shown below.

$1 - \frac{C}{C_{initial}}$	t (s)	$1 - \frac{C}{C_{initial}}$	t (s)
0.01	10,300	0.4	11,047
0.02	10,405	0.5	11,291
0.04	10,527	0.6	11,399
0.1	10,636	0.7	11,435
0.2	10,846	0.8	11,621
0.3	10,996	0.9	11,931

From this information calculate the molecular diffusion coefficient of $KMnO_4$ in water in this concentration range. Compare your results with literature values of

$$\mathfrak{D} = 4.35 \times 10^{-10} \text{ m}^2/\text{s to } 15 \times 10^{-10} \text{ m}^2/\text{s}.$$

8. Kerosene and gasoline are pumped successively at 1.1 m/s through a 25.5 cm ID pipeline 1,000 km long. Calculate the 5/95–95/5% contaminated width at the exit of the pipe given that the kinematic viscosity for the 50/50% mixture is

$$\mu/\rho = 0.9 \times 10^{-6} \text{ m}^2/\text{s}.$$

9. Water is drawn from a lake, flows through a pump, and passes down a long pipe in turbulent flow. A slug of tracer (not an ideal pulse input) enters the intake line at the lake and is recorded downstream at two locations in the pipe L meters apart.

The mean residence time of fluid between recording points is 100 s, and variance of the two recorded signals is

$$\sigma_1^2 = 800 \text{ s}^2,$$

$$\sigma_2^2 = 900 \text{ s}^2.$$

What would be the spread of an ideal pulse response for a section of this pipe, free from end effects and of length $L/5$?

References

Sjenitzer, Pipeline Engineer, December 1958
Petroleum Refiner 37 191 (March 1958)
Pipe Line Industry, pg 51 (May 1958)
Blackwell, *Local AIChE meeting*, Galveston, TX, October 18, 1957
Pratt et al. CES 28 1901 (1973)
Taylor, G.I. *Proc. Roy. Soc. (London)* **219A** 186 (1963)

Chapter 8
The Tanks-in-Series Model

This model can be used whenever the dispersion model is used; and for not too large a deviation from plug flow, both models give identical results, for all practical purposes. Which model you use depends on your mood and your likes and dislikes.

The dispersion model has the advantage that it is used in all correlations for flow in real tubes, pipes, and packed beds. On the other hand, the tanks-in-series model is simple, can be used with any kinetics, and it can be extended without too much difficulty to any arrangement of compartments, with or without recycle.

8.1 The RTD from Pulse Response Experiments and the Tanks-in Series Model

Let us define

$\theta_i = t/\bar{t}_i$ = dimensionless time based on the mean residence time per tank $\bar{t}_i$

$\theta = t/\bar{t}$ = dimensionless time based on the mean residence time for all N tanks, $\bar{t}$.

Then $\theta_i = N\theta \cdot$

The RTD is found to be

$$\text{for } N = 1 \ \ldots \ \bar{t}_i \mathbf{E} = e^{-t/\bar{t}_i}$$

$$\text{for } N = 2 \ \ldots \ \bar{t}_i \mathbf{E} = \frac{t}{\bar{t}_i} e^{-t/\bar{t}_i}$$

$$\text{for } N = 3 \ \ldots \ \bar{t}_i \mathbf{E} = \frac{1}{2} \left(\frac{t}{\bar{t}_i} \right)^2 e^{-t/\bar{t}_i} \ \ldots \ \text{and so on}.$$

$$\vdots$$

$$\text{for } N \ \ldots \ \bar{t}_i \mathbf{E} = \frac{1}{(N-1)!} \left(\frac{t}{\bar{t}_i} \right)^{N-1} e^{-t/\bar{t}_i}$$

O. Levenspiel, *Tracer Technology*, Fluid Mechanics and Its Applications 96, DOI 10.1007/978-1-4419-8074-8_8, © Springer Science+Business Media, LLC 2012

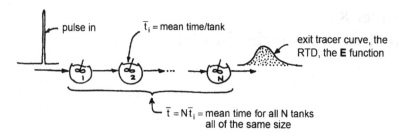

Fig. 8.1

For the number of tanks N, the RTD, mean, and variance for this model are (MacMullin and Weber 1935):

$$\bar{t}\mathbf{E} = \left(\frac{t}{\bar{t}}\right)^{N-1}\frac{N^N}{(N-1)!}e^{-tN/\bar{t}} \quad \dots \bar{t} = N\bar{t}_i \quad \dots \sigma^2 = \frac{\bar{t}^2}{N}$$

$$\bar{t}_i\,\mathbf{E} = \left(\frac{t}{\bar{t}_i}\right)^{N-1}\frac{1}{(N-1)!}e^{-t/\bar{t}_i} \quad \dots \bar{t}_i = \frac{\bar{t}}{N} \quad \dots \sigma^2 = N\bar{t}_i^{\;2} = \frac{\bar{t}^2}{N}$$

$$\mathbf{E}_{\theta i} = \bar{t}_i\mathbf{E} = \frac{\theta_i^{N-1}}{(N-1)!}e^{-\theta_i} \quad \dots \bar{t}_{\theta i} = N \dots \sigma^2_{\theta i} = N$$

$$\mathbf{E}_\theta = (N\bar{t}_i)\mathbf{E} = \frac{N(N\theta)^{N-1}}{(N-1)!}e^{-N\theta} \dots \bar{t}_\theta = 1 \dots \sigma^2_\theta = \frac{1}{N}$$

(8.1)

This is shown graphically in Figs. 8.1 and 8.2.

The tracer curves for the different representations and their relationships for three tanks in a series, $N = 3$, are shown in Fig. 8.3.

The properties of the RTD curves are sketched in Fig. 8.3. For $N > 50$ the RTD becomes just about symmetrical and Gaussian, see Fig. 8.4.

8.1.1   Comments

(a) *For small deviation from plug flow, $N > 50$, the RTD becomes symmetrical and Gaussian. Thus, (8.1) can be approximated by (Fig. 8.5)*

$$\mathbf{E}_\theta = \frac{1}{\sqrt{2\pi\sigma^2}}\exp\left[-\frac{(1-\theta)^2}{2\sigma^2}\right] \dots \text{ the Gaussian distribution}$$

$$= \sqrt{\frac{N}{2\pi}}\exp\left[-\frac{(1-\theta)^2}{2/N}\right] = \sqrt{\frac{N}{2\pi}}\exp\left[-\frac{(1-t/\bar{t})^2}{2/N}\right] \dots \text{ close to Gaussian}$$

(8.2)

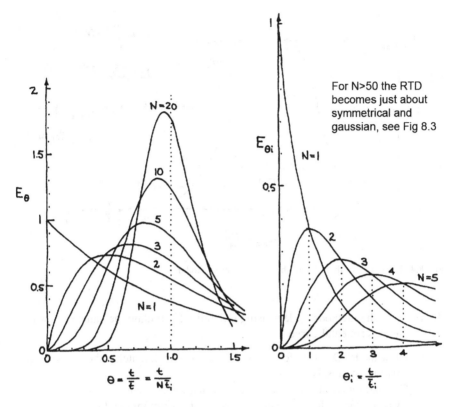

Fig. 8.2   E curves for the tanks-in-series model

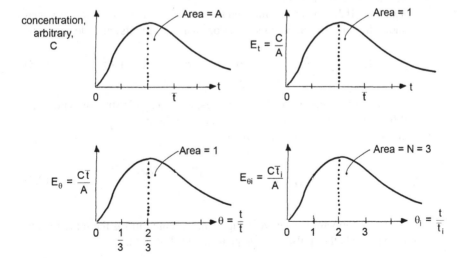

Fig. 8.3   RTD for $N = 3$

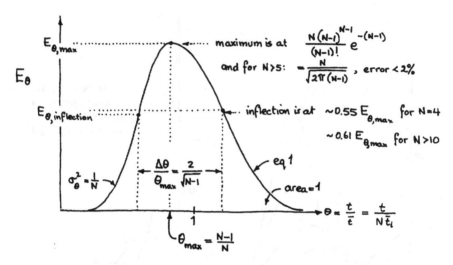

Fig. 8.4 Properties of the RTD of the tanks-in-series model

(b) Here are the methods of decreasing reliability for finding the value of N which fits an experimental curve

- Draw the RTD curves for various N and see which matches the experimental curve most closely
- Calculate σ^2 from experiment and compare with theory
- Evaluate the width of the curve at 61% of maximum height
- Match the maximum height

and there are still other ways.

(c) *Independence.* If M tanks are connected to N more tanks (all of same size), then their individual means and variances (in ordinary time units) are additive, or

$$\bar{t}_{M+N} = \bar{t}_M + \bar{t}_N \quad \dots \quad \text{and} \quad \dots \quad \sigma^2_{M+N} = \sigma^2_M + \sigma^2_N. \qquad (8.3)$$

Because of this property, we can join streams and recycle streams. Thus, this model becomes useful for treating recirculating systems.

(d) For any *one-shot tracer* input to a vessel

$$\Delta\sigma^2 = \sigma^2_{out} - \sigma^2_{in} = \frac{(\bar{t}_{out} - \bar{t}_{in})^2}{N} = \frac{(\Delta\bar{t})^2}{N}. \qquad (8.4)$$

The increase in variance between input and output points is the same for any kind of input, whether pulse, double peaked, etc (Fig. 8.6).

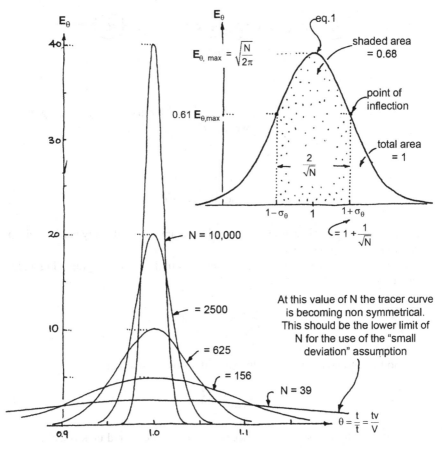

Fig. 8.5 RTD for small deviations from plug flow. This figure is identical to Fig. 6.3 where $2(D/uL)$ is replaced by $1/N$

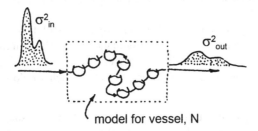

Fig. 8.6 For a one-shot input

(e) Relationship between the dispersion and the tanks-in-series models. From (6.1) or (8.1)

$$\sigma_\theta^2 = 2\left(\frac{D}{uL}\right) = \frac{1}{N} \quad \text{or} \quad \sigma^2 = \frac{\bar{t}^2}{N}. \tag{8.5}$$

For large deviation from plug flow, see Levenspiel (1962).

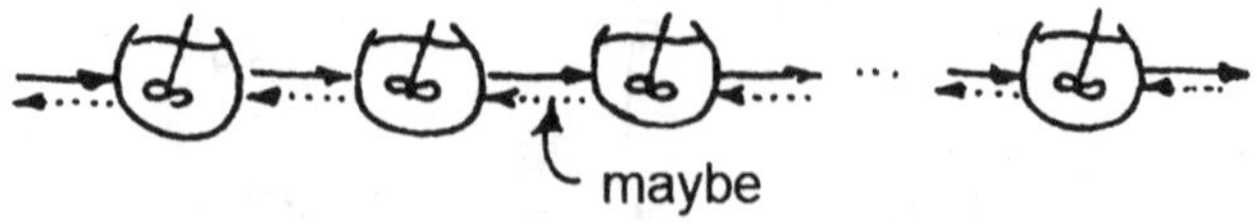

Fig. 8.7 A more complex model

(f) *Value of N for flow of gases in fixed beds.* From the fixed bed graph, Fig. 6.22, where u = mean axial velocity, we find $D\epsilon/u_o d_p \cong D/u d_p \cong \frac{1}{2}$.
With (8.5) we get

$$N_{\text{tanks}} \cong \left(\begin{array}{c} n \text{ particles end to end} \\ \text{in the packed bed} \end{array} \right) = \frac{L}{d_p}. \qquad (8.6)$$

Thus, each particle acts as one stirred tank, and a bed of many particles deep very closely approaches plug flow.

(g) *Value of N for flow of liquids in fixed beds.* From the fixed bed graph, Fig. 6.22, we find

$$\frac{\mathbf{D}\epsilon}{u_o d_p} = \frac{\mathbf{D}}{u d_p} \cong 2.$$

So by arguments similar to the case of gases we find that

$$N_{\text{tanks}} \cong \frac{L}{4d}. \qquad (8.7)$$

Thus, each four particles in succession act as one stirred tank (ideal).

(h) *The boundary conditions.* Here we have no problem as with the dispersion model. The entrance and exit are usually well approximated by the closed vessel boundary conditions, so the measured $\mathbf{E}$ and $\mathbf{F}$ curves are in fact the proper $\mathbf{E}$ and $\mathbf{F}$ curves.

Whatever the boundary conditions, we must still take note of how the measurements are taken ... either instantaneously or by "mixing cup."

If you do have backmixing or dispersion at inlet and outlet, and also between tanks, then you get a different model which we do not consider here (Fig. 8.7).

(i) Before deciding to use this model be sure to check the shape of the experimental curve to see if the model really applies. Do not use the model indiscriminately.

8.2 Tanks-in-Series, with Recycle and Dead Spaces

Here we combine this chapter with the compartment models as shown in Fig. 8.8.

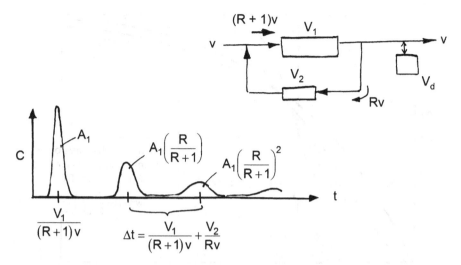

Fig. 8.8 Tanks-in-series with recycle and dead spaces

8.3   Tanks-in-Series: F Curve

The output **F** curve from a series of N ideal-stirred tanks is, in its various forms, given by (8.8).

$$\mathbf{F} = 1 - e^{-N\theta}\left[1 + N\theta + \frac{(N\theta)^2}{2!} + \cdots + \frac{(N\theta)^{N-1}}{(N-1)!} + \cdots\right],$$

$$\mathbf{F} = 1 - e^{-\theta_i}\left[1 + \theta_i + \frac{\theta_i^2}{2!} + \cdots + \frac{\theta_i^{N-1}}{(N-1)!} + \cdots\right],$$

$$\text{Number of tanks}\begin{cases} \text{For } N = 1 \text{ use the first term} \\ \text{For } N = 2 \text{ use two terms} \\ \text{For } N = 3 \text{ use three terms} \\ \text{For } N \text{ tanks, use } N \text{ terms} \end{cases}. \qquad (8.8)$$

This is shown in graphical form in Fig. 8.9.

Example 1. *Modifications to a Winery.* A small diameter pipe 32 m long runs from the fermentation room of a winery to the bottle filling room. Sometimes red wine is pumped through the pipe, sometimes white, and whenever the switch is made from one to the other a small amount of "house blend" rosé is produced (eight bottles).

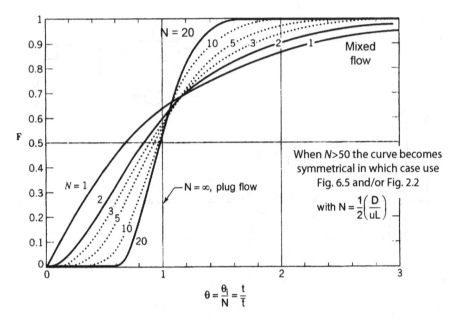

Fig. 8.9   The F curve for the tanks-in-series model, from MacMullin and Weber (1935)

Because of some construction in the winery, the pipeline length will have to be increased to 50 m. For the same flow rate of wine, how many bottles of rosé may we now expect to get each time we switch the flow?

Solution. Figure 8.10 sketches the problem. Let the number of bottles of rosé be related to σ.

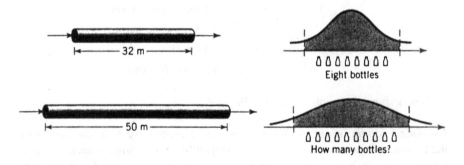

Fig. 8.10   Before and After

Original: $N_1 = 8$ $L_1 = 32$ m $\sigma_1 = 8$ $\sigma_1^2 = 64$

Longer pipe: $N_2 = ?$ $L_2 = 50$ m $\sigma_2 = ?$ $\sigma_2^2 = ?$

Then from (8.1) $\sigma^2 \propto N$ or $\sigma^2 \propto L$.

$$\therefore \frac{\sigma_2^2}{\sigma_1^2} = \frac{L_2}{L_1} = \frac{50}{32}$$

$$\therefore \sigma_2^2 = \frac{50}{32}(64) = 100$$

$\therefore \sigma_2 = 10$ ... or we can expect <u>10 bottles of *vin rosé*</u>.

***Example 2**. A Fable on River Pollution.* Last spring our office received complaints of a large fish kill along the Ohio River, indicating that someone had discharged highly toxic material into the river. Our water monitoring stations at Cincinnati and at Portsmouth, Ohio (119 miles apart), report that a large slug of phenol was moving down the river, and we strongly suspect that this is the cause of the pollution. The slug took about 10.5 h to pass the Portsmouth monitoring station, and its concentration peaked at 8:00 A.M. Monday. About 26 h later the slug peaked at Cincinnati, taking 14 h to pass this monitoring station.

Phenol is used at a number of locations on the Ohio River, and their distances upriver from Cincinnati are as follows:

Ashland, KY	150 miles upstream	Marietta, OH	303
Huntington, WV	168	Wheeling, WV	385
Pomeroy, OH	222	Steubenville, OH	425
Parkersburg, WV	290	Pittsburgh, PA	500

What can you say about the probable pollution source?

Solution. Let us first sketch what is known, as shown in Fig. 8.11.

To start, assume that a perfect pulse is injected. Then according to any reasonable flow model, either dispersion or tanks-in-series, we have

$$\sigma_{\text{tracer curve}}^2 \propto \left(\begin{array}{c} \text{distance from} \\ \text{point of origin} \end{array} \right)$$

or

$$\left(\begin{array}{c} \text{spread of} \\ \text{curve} \end{array} \right) \propto \sqrt{ \begin{array}{c} \text{distance from} \\ \text{origin} \end{array} }$$

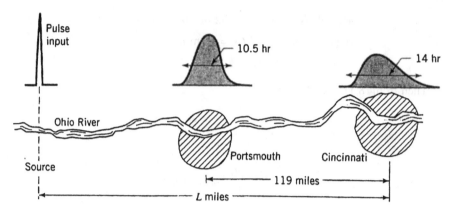

Fig. 8.11

$$\therefore \text{ from Cincinnati} \quad 14 = kL^{1/2}$$
$$\left.\therefore \text{ from Portsmouth} \quad 10.5 = k(L - 119)^{1/2}\right\}$$

Dividing one by the other gives

$$\frac{14}{10.5} = \sqrt{\frac{L}{L - 119}} \ldots \text{ from which } L = 272 \text{ miles.}$$

Comment. Since the dumping of the toxic phenol may not have occurred instantaneously, any location where $L \leq 272$ miles is suspect, or

$$\left.\begin{array}{c} \text{Ashland} \\ \text{Huntington} \\ \text{Pomeroy} \end{array}\right\} \leftarrow$$

This solution assumes that different stretches of the Ohio River have the same flow and dispersion characteristics (reasonable), and that no suspect tributary joins the Ohio within 272 miles of Cincinnati. This is a poor assumption ... check a map for the location of Charleston, WV, on the Kanawah River which runs into the Ohio River.

***Example 3**. Finding the Vessel E Curve Using a Sloppy Tracer Input.* Given C_{in} and C_{out} as well as the location and spread of these tracer curves, as shown in Fig. 8.12, estimate the vessel **E** curve. We suspect that the tanks-in-series model reasonably represents the flow through the vessel.

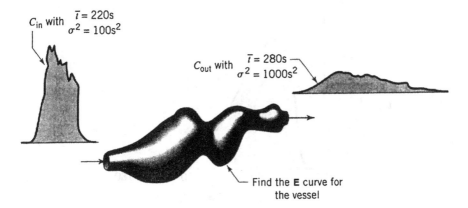

Fig. 8.12

Solution. From Fig. 8.12 we have, for the vessel,

$$\Delta \bar{t} = 280 - 220 = 60 \text{ s},$$

$$\Delta(\sigma^2) = 1,000 - 100 = 900 \text{ s}^2.$$

Equation (8.3) represents the tanks-in-series model and gives

$$N = \frac{(\Delta \bar{t})^2}{\Delta(\sigma^2)} = \frac{60^2}{900} = 4 \text{ tanks.}$$

So from (8.1), for N tanks-in-series we have

$$E = \frac{t^{N-1}}{\bar{t}^N} \frac{N^N}{(N-1)!} e^{-tN/\bar{t}}$$

and for $N = 4$

$$\mathbf{E} = \frac{t^3}{60^4} \frac{4^4}{3 \times 2} e^{-4t/60},$$

or $\underline{\underline{\mathbf{E} = 3.2922 \times 10^{-6} t^3 e^{-0.0667t}}}.$

Figure 8.13 shows the shape of this **E** curve.

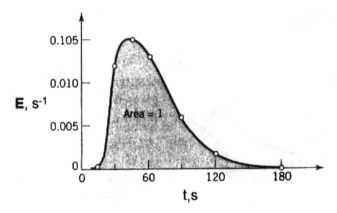

Fig. 8.13 E curve for vessel

Example 4. *Storage Tank for Radioactive Waste.* At Oak Ridge National Laboratories, strongly radioactive waste fluids are stored in "safe-tanks" which are simply long, small-diameter (e.g., 20 m by 10 cm) slightly sloping pipes. To avoid sedimentation and development of "hot spots," and also to insure uniformity before sampling the contents, fluid is recirculated in these pipes.

To model the flow in these tanks, a pulse of tracer is introduced and the curve of Fig. 8.14 is recorded. With pencil and ruler, develop a suitable model for this system and evaluate its parameters.

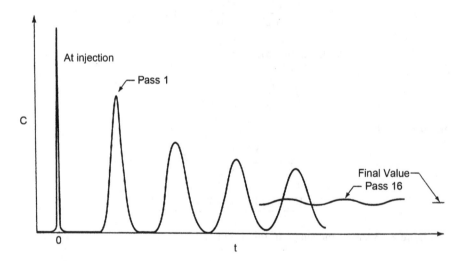

Fig. 8.14 RTD for a closed recirculating system

Solution. Try the tanks-in-series model. Then with ruler measure the width and location of the peaks of the curves. This gives Fig. 8.15.

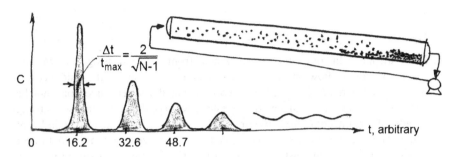

Fig. 8.15

From the tracer curve, Fig. 8.15, we have

$$N = 1 + 4\left(\frac{t_{max}}{\Delta t}\right)^2$$

from which we find

First pass	$N = 1 + 4(t_{max}/\Delta t)^2 = 1 + 4\,(16.2/3.2)^2 = 102$ tanks	
Second pass	$N = 202$	
Third pass	$N = 304$	$\therefore N \cong 100$ tanks

Problems

1. From a pulse input into a vessel we obtain the following output signal

Time (min)	1	3	5	7	9	11	13	15
Concentration (arbitrary)	0	0	10	10	10	10	0	0

We want to represent the flow through the vessel with the tanks-in-series model. Determine the number of tanks to use.

2. Fit the tanks-in-series model to the following output data to a pulse input.

(a) Use linear interpolation between the data points.
(b) Use something better than linear interpolation.

T	0	2	4	6	8	10	12
C	0	2	10	8	4	2	0

The concentrations are instantaneous readings taken at the exit of the vessel.

3. Fit the tanks-in-series model to the following mixing cup output data to a pulse input.

(a) Use linear interpolation between the data points.
(b) Use something better than linear interpolation.

T	0–2	2–4	4–6	6–8	8–10	10–12
C	2	10	8	4	2	0

4. Denmark has a factory which extracts agar from seaweed using a seven tank extraction unit. In flowing through the system the viscosity of the fluid changes over tenfold and the engineers are worried that the seven tanks may not be behaving ideally. To put this worry to rest the RTD of the unit is measured using a tracer of 200 mCi of radioactive Br-82 in the form of NH_4Br dissolved in 1 L of water. The tracer is dumped into the first tank at time zero and the output tracer concentration from the last tank is recorded. How many ideal stirred tanks in series represent the real system?

t (min)	C (arbitrary)	t (min)	C (arbitrary)
180	0	1,006	242
206	4	1,126	214
266	12	1,360	118
296	42	1,600	78
416	88	1,836	42
534	160	2,070	18
650	216	2,540	10
770	244	3,000	(0)
890	254		

Data from the Danish Isotope Center, report of May 1977.

5. In physiological studies the tracer injection and sampling system (ISS) may not be small compared to the organism being studied, for example for hummingbirds. In such situations, it may be important to correct for the time delay and distortion introduced by imperfect pulse injection and by the lines leading to and from the ISS.

    In one such study, with indocyanine green dye in blood (0.038 mg/mL) as flow tracer, we connect the injection needle directly to the mouth of the sampling line, we introduce a standard pulse at $t = 0$, and we record its response (curve A). We then repeat this procedure with the injection needle and sampling probe in the organism to be studied (curve B). Results of a typical run are as follows (Fig. 8.16):

    a) In representing the ISS by the tanks-in-series model, what values of parameters should we use?

    b) What flow model would you use for the organism?

    Data from Oleksy et al. (1969).

6. Our 16 stirred tank system (1 m³ each) is contaminated, consequently we have to flush it with fresh water (1 m³/min) so that nowhere in the system is the concentration of contaminant more than 0.1% its present value. How long must we flush

Fig. 8.16 From Oleksy
(1969)

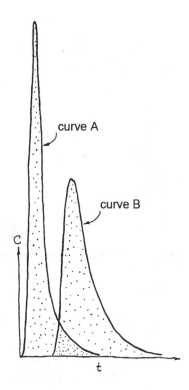

curve A

curve B

C

t

(a) If each tank is fed fresh water and discharges its own waste?
(b) If the tanks are connected in series, the first tank is fed fresh water, and only
   the last tank discharges waste?
7. Fit the RTD of Fig. 8.17 with the tanks-in-series model.
8. Gas and liquid pass concurrently upward through a vertical reactor packed with
   catalyst ($V_{total} = 5$ m^3, fraction solids $= 0.64$). A pulse test on the liquid

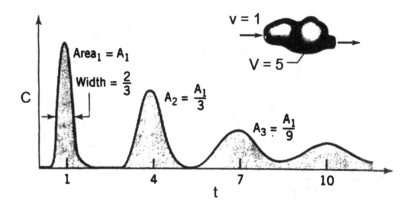

$v = 1$

$V = 5$

Area$_1$ = A$_1$

Width = $\frac{2}{3}$

C

$A_2 = \frac{A_1}{3}$

$A_3 = \frac{A_1}{9}$

1   4   7   10

t

Fig. 8.17

($v_L = 2,000$ L/min) gives the tracer curve of Fig. 8.18. Develop a tank-in-series flow model to represent the flow of liquid through the column.

9. In Mark Kurlansky's book "SALT, a World History," pgs. 82–83, *Penguin Books*, 2002, we read

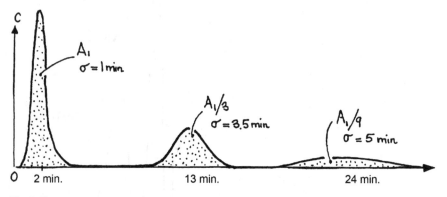

Fig. 8.18

Between the sixth and ninth centuries, the last great technical advances in salt manufacturing until the twentieth century was invented. Instead of trapping sea water in a single artificial pond, closing it off, and waiting for the sun to evaporate the water the salt makers built a series of ponds. The first, a large open tank had a system of pumps and sluices that moved the pond's seawater to the next pond ... and so on.

What do you think of this great technical advance? What are its advantages, disadvantages, etc.? Discuss please and see Fig. 8.19.

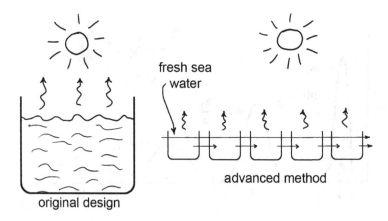

Fig. 8.19

References

Levenspiel O. Comparison of the tanks-in-series and the dispersion models for the non-ideal flow of fluid. CES 17 576 (1962)

Oleksy et al. *J. App. Physiology* 26 227 (1969)

MacMullin & Weber. *Trans AIChE* 31 409 (1935)

Chapter 9
Convection Model for Laminar Flow in Pipes

When a vessel is long enough, then the dispersion or tanks-in-series models well describe its flow behavior. But how long is "long enough?" For packed beds and for turbulent flow in pipes, just about any vessel length is long enough; however, for laminar flow in pipes we may be in other flow regimes, in particular, that of the pure convection model. But first let us look at the extremes for laminar flow

- If the tube is long enough, then molecular diffusion in the lateral direction will have enough time to distort the parabolic velocity profile so that the *dispersion model* applies.
- If the tube is short and the flow rate high, then molecular diffusion has not enough time to act so all we need to consider as causing a spread in residence time of fluid is the velocity profile. The flow is then in the *pure convection* regime.
- If flow is so slow that the main movement of fluid is by molecular diffusion, not by bulk flow, then we enter the *pure diffusion* regime. We rarely meet this situation outside of reservoir engineering.

Figure 9.1 shows RTD curves typical of these regimes.

Note how very different are these RTD curves for flow of Newtonians in circular pipes.

Figure 9.2 tells what regime you are in and which model to use. Just locate the point on the chart which corresponds to the fluid being used (Schmidt number), the flow conditions (Reynolds number), and vessel geometry (L/d_t). But be sure to check that you are not in turbulent flow. This chart only has meaning if you have laminar flow.

$\mathcal{D}/ud_t$ is the reciprocal of the *Bodenstein number*. It measures the mixing contribution made by molecular diffusion. It is *not* the axial dispersion number, $\mathbf{D}/ud_t$, except in the pure diffusion regime. Figure 9.1 is adapted from Ananthakrishnan et al. (1965). See Fig. 9.1 for approximate and numerical solutions for the intermediate regimes.

O. Levenspiel, *Tracer Technology*, Fluid Mechanics and Its Applications 96,
DOI 10.1007/978-1-4419-8074-8_9, © Springer Science+Business Media, LLC 2012

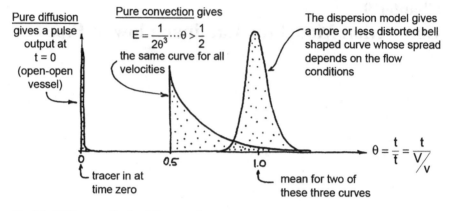

Fig. 9.1 RTD curves for the different regimes

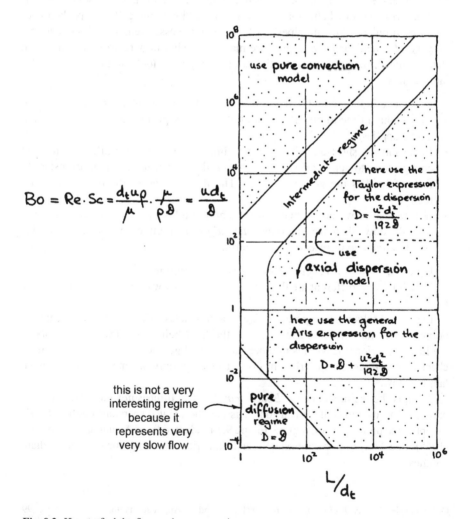

Fig. 9.2 How to find the flow regime you are in

Gases are likely to be in the dispersion regime, not the pure convection regime. Liquids can well be in one regime or other. Very viscous liquids, such as polymers, are likely to be in the pure convection regime. If your system falls in the no-man's-land between regimes, calculate the vessel behavior based on the two bounding regimes and then try averaging because the numerical solution is impractically complex to use.

The *pure convection model* assumes that each element of fluid slides past its neighbor with no interaction by molecular diffusion. Thus, the spread in residence times is caused only by velocity variations.

This chapter deals with this model (Fig. 9.3).

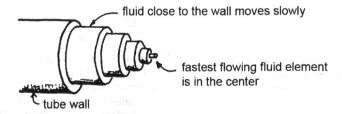

Fig. 9.3 Pipe flows for pure convection regimes

9.1   Pulse Response Experiment for Newtonians and the E Curve for Flow in Pipes

The shape of the response curve is extremely influenced by the way tracer is introduced into the flowing fluid, and how it is measured. You may inject it in two distinctly different ways, and you can measure it in two distinctly different ways, see Fig. 9.4.

We therefore have four combinations of boundary conditions, each with its own particular **E** curve (Fig. 9.5).

It can be shown that

- **E** is the proper response curve. It is the curve treated in the previous chapters. It represents the RTD in the vessel.
- **E*** or ***E** is identical always, so we will call them **E*** from now on. One correction for this planar b.c. will bring or transform this to the proper RTD.
- **E**** requires two corrections, one for the entrance and the other for the exit, to transform it to a proper RTD.

It may be simpler experimentally to determine **E*** or **E**** rather than **E**. This is perfectly all right. However, remember to transform these measured tracer curves to the **E** curve before calling them the RTD.

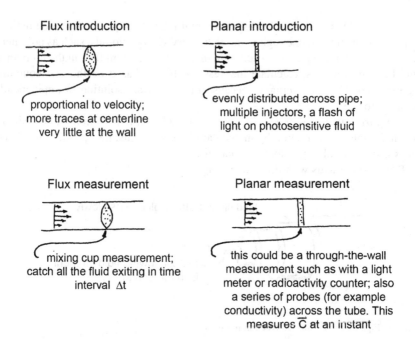

Flux introduction

proportional to velocity;
more traces at centerline
very little at the wall

Planar introduction

evenly distributed across pipe;
multiple injectors, a flash of
light on photosensitive fluid

Flux measurement

mixing cup measurement;
catch all the fluid exiting in time
interval Δt

Planar measurement

this could be a through-the-wall
measurement such as with a light
meter or radioactivity counter; also
a series of probes (for example
conductivity) across the tube. This
measures $\overline{C}$ at an instant

Fig. 9.4 Various ways to introduce and measure tracer

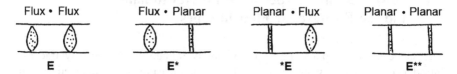

Flux • Flux   Flux • Planar   Planar • Flux   Planar • Planar

E   E*   *E   E**

Fig. 9.5 Four combinations of input and output and their **E** curves

For Newtonians flowing in circular pipes and tubes the various pulse response curves are found to be as follows:

$$\mathbf{E} = \frac{\bar{t}^2}{2t^3} \quad \text{for } t \geq \frac{\bar{t}}{2} \quad \text{and} \quad \mu = \bar{t} = \frac{V}{v},$$

$$\mathbf{E}_\theta = \frac{1}{2\theta^3} \quad \text{for } \theta \geq \frac{1}{2} \quad \text{and} \quad \mu_\theta = 1,$$

$$\left.\begin{aligned}\mathbf{E}^* &= \frac{\bar{t}}{2t^2} \quad \text{for } t \geq \frac{\bar{t}}{2} \\[6pt] \mathbf{E}_\theta^* &= \frac{1}{2\theta^2} \quad \text{for } \theta \geq \frac{1}{2}\end{aligned}\right\} \text{ and } \mu^* = \infty, \qquad (9.1)$$

$$\left.\begin{aligned}\mathbf{E}^{**} &= \frac{1}{2t} \quad \text{for } t \geq \frac{\bar{t}}{2} \\[6pt] \mathbf{E}_\theta^{**} &= \frac{1}{2\theta} \quad \text{for } \theta \geq \frac{1}{2}\end{aligned}\right\} \text{ and } \mu^{**} = \infty.$$

These equations are shown graphically in Fig. 9.6.

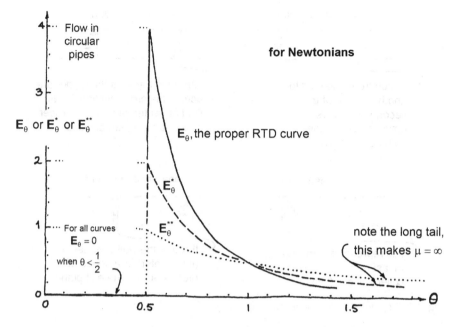

Fig. 9.6 E curves for laminar flow of Newtonians in circular pipes

Note the simple relationship between $\mathbf{E}$, $\mathbf{E}^*$, and $\mathbf{E}^{**}$. Thus, at any time we can write

Note that Eq (2) applies for flow through circular pipes.   or

$$E_\theta^{**} = \theta E_\theta^* = \theta^2 E_\theta$$

$$E^{**} = \frac{t}{\bar{t}}E^* = \frac{t^2}{\bar{t}^2}E$$

$$\bar{t} = V/v$$

(9.2)

9.2   Step Response Experiment and the F Curve for Newtonians

The way tracer is introduced and the way it is measured will determine the shape of the resulting curve. Again we have two main ways of introducing and measuring tracer (Fig. 9.7).

Thus, again, we have four combinations of boundary condition with their corresponding **F** curves (Fig. 9.8).

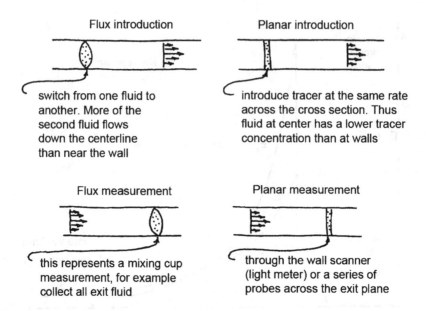

Fig. 9.7 Ways to introduce and measure step tracer

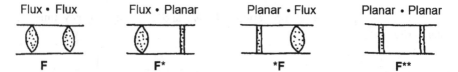

Fig. 9.8 Four different **F** curves

Again $\mathbf{F}^* = {}^*\mathbf{F}$, and $\mathbf{F}$ is the proper step response function. The equations shapes and relationships between these $\mathbf{F}$ curves are as follows.

$$\mathbf{F} = 1 - \frac{1}{4\theta^2} \quad \text{for } \theta \geq \frac{1}{2},$$

$$\mathbf{F}^* = 1 - \frac{1}{2\theta} \quad \text{for } \theta \geq \frac{1}{2},$$

$$\mathbf{F}^{**} = \frac{1}{2}\ell n 2\theta \quad \text{for } \theta \geq \frac{1}{2}. \tag{9.3}$$

Graphically, these equations for flow in circular pipes are shown in Fig. 9.9.

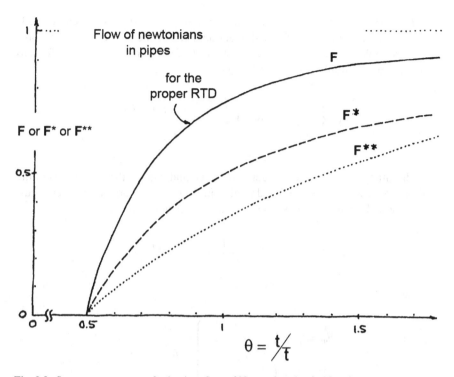

Fig. 9.9 Step response curves for laminar flow of Newtonians in circular pipes

Also, each **F** curve is related to its corresponding **E** curve. Thus, at any time t_1 or θ_1

$$\mathbf{F}^* = \int_0^{t_1} \mathbf{E}_t^* \, dt = \int_0^{\theta_1} \mathbf{E}_\theta^* \, d\theta \quad \text{and} \quad \mathbf{E}_t^* = \left.\frac{d\mathbf{F}^*}{dt}\right|_{t_1} \quad \text{or} \quad \mathbf{E}_\theta^* = \left.\frac{d\mathbf{F}^*}{d\theta}\right|_{\theta_1}. \qquad (9.4)$$

Similarly for the relationship between $\mathbf{E}^{**}$ and $\mathbf{F}^{**}$, and between $\mathbf{E}$ and $\mathbf{F}$.

9.2.1 Comments

(a) *Test for an RTD curve.* Proper RTD curves must satisfy the material balance checks (zero and first moment).

$$\int_0^\infty \mathbf{E}_\theta \, d\theta = 1 \quad \text{and} \quad \int_0^\infty \theta \mathbf{E}_\theta \, d\theta = 1.$$

The **E** curves of this chapter, for non-Newtonians and all shapes of channels, all meet this requirement. All the **E*** and **E**** curves of this chapter do not, however.

(b) *RTD curves from experiment.* If you wish to represent data by an empirical **E** curve, be sure that the empirical equation satisfies the above two conditions. For example, in the literature we sometimes find the following equation form to represent coiled tubes, screw extruders, and the like

$$
\mathbf{E}_\theta \begin{cases} = \dfrac{AB}{\theta^{B+1}} \\ = 0 \end{cases} \quad \text{or} \quad \mathbf{F} \begin{cases} = \dfrac{A}{\theta^B} & \text{for } \theta \geq \theta_0, \\ = 0 & \text{for } \theta < \theta_0. \end{cases}
$$

In satisfying the above two conditions, we find that the three constants of the above expression are interrelated to give just one independent parameter. Thus, if we take the breakthrough time θ_0 as the independent measure then

$$
A = \theta_0^{1/(1-\theta_0)}, \quad B = \frac{1}{1-\theta_0}, \quad A = \left(\frac{B-1}{B}\right)^B
$$

and

$$
\left. \begin{aligned} \mathbf{E}_\theta &= \frac{1}{1-\theta_0} \frac{1}{\theta} \left(\frac{\theta_0}{\theta}\right)^{1/(1-\theta_0)} \\ \mathbf{F} &= 1 - \left(\frac{\theta_0}{\theta}\right)^{1/(1-\theta_0)} \end{aligned} \right\} \quad \text{for } \theta > \theta_0.
$$

The same sort of argument holds for other types of RTD curves. This means that the parameters can not all be independently chosen.

(c) *The variance and other RTD descriptors.* The variance of all the **E** curves of this chapter is finite; but it is infinite for all the **E*** and **E**** curves. So be sure you know which curve you are dealing with.

In general the convection model **E** curve has a long tail. This makes the measurement of its variance unreliable; thus is not a useful parameter for convection models and is not presented here.

The breakthrough time θ_0 is probably the most reliably measured and most useful descriptive parameter for convection models, so it is widely used.

9.3   Non-Newtonians and Noncircular Channels

(a) Power law fluids in pipes
The shear force vs. velocity gradient relationship and the resulting velocity profile for power law fluids are (Fig. 9.10)

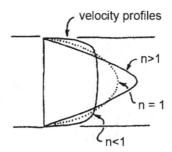

velocity profiles

n>1

n = 1

n<1

Fig. 9.10 Power law fluids

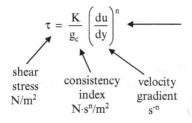

$$\tau = \frac{K}{g_c}\left(\frac{du}{dy}\right)^n$$

shear stress N/m^2

consistency index $N{\cdot}s^n/m^2$

velocity gradient s^{-n}

where n is called the behavior index

n ≅ 0.4 ~ 0.6 for many foods: soups, sauces, tomato juice

n ≅ 0.1 ~ 0.2 for slurries of fine solids in water: lime water, cement rock in water, clay in water

Solving the equations for the pulse response curve gives

$$\left.\begin{array}{l} \mathbf{E}_\theta = \dfrac{2n}{3n+1}\dfrac{1}{\theta^3}\left[1 - \dfrac{n+1}{3n+1}\dfrac{1}{\theta}\right]^{n-1/n+1} \quad \text{for } \theta \ge \dfrac{n+1}{3n+1} \\[4mm] \mathbf{E}_\theta^* = \theta\cdot\mathbf{E}_\theta \quad \text{and} \quad \mathbf{E}_\theta^{**} = \theta^2\mathbf{E}_\theta \end{array}\right\} \tag{9.5}$$

and for the step response, following (9.4),

$$\left.\begin{array}{l} \mathbf{F} = \left[1 - \dfrac{n+1}{3n+1}\dfrac{1}{\theta}\right]^{2n/n+1}\left[1 + \dfrac{2n}{3n+1}\dfrac{1}{\theta}\right] \quad \text{for } \theta \ge \dfrac{n+1}{3n+1} \\[4mm] \mathbf{F}^* = \left[1 - \dfrac{n+1}{3n+1}\dfrac{1}{\theta}\right]^{2n/n+1} \qquad\qquad\qquad \text{for } \theta \ge \dfrac{n+1}{3n+1} \\[4mm] \mathbf{F}^{**} \text{ gives an awkward expression} \end{array}\right\}. \tag{9.6}$$

Note that (9.4) relates each **E** to its corresponding **F** expression.
Figures 9.11 and 9.12 are the graphical representation for various *n* values.

(b) Bingham plastics in pipes
The shear stress vs. velocity gradient relationship is (Fig. 9.13)

$$\tau = \tau_0 + \left(\frac{n}{g_c}\right)\left(\frac{du}{dy}\right)$$

yield stress

plastic viscosity

- For butter just out of the refrigerator: $\tau_0 = 100 \sim 100 \text{ N/m}^2$
- For warm just ready to melt butter: $\tau_0 = 10–20 \text{ N/m}^2$

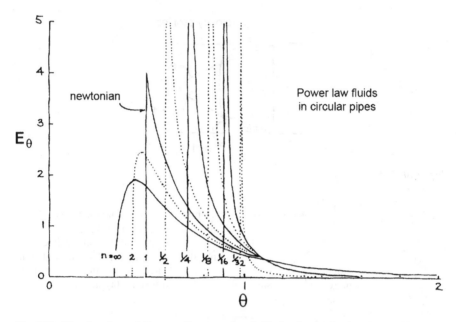

Fig. 9.11 The closed vessel **E** curves for power law fluids flowing in circular pipes

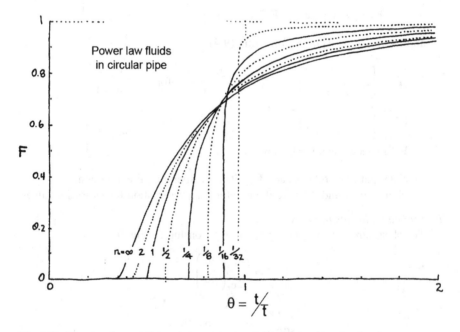

Fig. 9.12 The closed vessel **F** curves for power law fluids flowing in a circular pipes

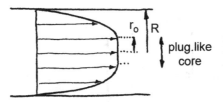

Fig. 9.13 Velocity profile of Bingham plastics flowing in circular pipes

Going through the necessary mathematics, we find the following expressions for the pulse and step responses

$$\mathbf{E}_\theta = \frac{(1-m)\theta_o}{\theta^3}\left[1 - m + \frac{m}{\sqrt{1 - (\theta_o/\theta)}}\right] \quad \text{for } \theta \geq \theta_o \left.\begin{array}{c}\\\\\end{array}\right\},$$

$$\mathbf{E}_\theta^* = \theta\mathbf{E}_\theta \quad \text{and} \quad \mathbf{E}_\theta^{**} = \theta^2\mathbf{E}_\theta \tag{9.7}$$

$$\mathbf{F} = \frac{m^2}{\theta_o} + \frac{2(1-m)}{\theta_o}\left\{\frac{m}{3}\left(1 - \frac{\theta_o}{\theta}\right)^{1/2}\left(2 + \frac{\theta_o}{\theta}\right) + \frac{1-m}{4}\left[1 - \left(\frac{\theta_o}{\theta}\right)^2\right]\right\}$$

$$\mathbf{F}^* = \left[m + (1-m)\left(1 - \frac{\theta_o}{\theta}\right)^{1/2}\right]^2 \left.\begin{array}{c}\\\\\\\end{array}\right\}, \tag{9.8}$$

$$\mathbf{F}^{**} = \text{very complicated}$$

where

$$m = \frac{r}{R_o} = \frac{\tau_o}{\tau_{wall}} = \frac{4\tau_o L}{(\Delta p_{loss})d_t} = \; < 1$$

$$\theta_o = \left(\begin{array}{c}\text{residence time of the central}\\\text{plug of fast moving fluid}\end{array}\right) = \frac{m^2 + 2m + 3}{6}.$$

Graphically (Fig. 9.14)
For falling film flow (Fig. 9.15)

$$\mathbf{E}_\theta = \frac{1}{3\theta^3}\sqrt{1 - \frac{1}{3\theta}} \quad \text{for } \theta \geq \frac{2}{3} \left.\begin{array}{c}\\\\\end{array}\right\},$$

$$\mathbf{E}_\theta^* = \theta\mathbf{E}_\theta \quad \text{and} \quad \mathbf{E}_\theta^{**} = \theta^2\mathbf{E}_\theta \tag{9.9}$$

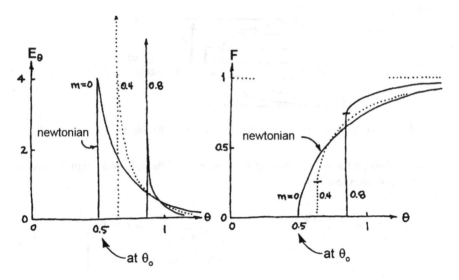

Fig. 9.14 E and F curves for Bingham plastics

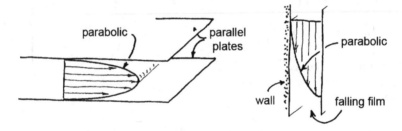

Fig. 9.15 Velocity profile for falling film flow

(c) Falling film flow of Newtonians or flow between parallel plates

$$\left.\begin{aligned}
\mathbf{F} &= \left(1 + \frac{1}{3\theta}\right)\sqrt{1 - \frac{2}{3\theta}} &&\text{for } \theta \geq \frac{2}{3} \\[2ex]
\mathbf{F}^* &= \sqrt{1 - \frac{2}{3\theta}} &&\text{for } \theta \geq \frac{2}{3} \\[2ex]
\mathbf{F}^{**} &= \frac{1}{3}\ell n\left[\frac{1 + \sqrt{1 - \dfrac{2}{3\theta}}}{1 - \sqrt{1 - \dfrac{2}{3\theta}}}\right] &&\text{for } \theta \geq \frac{2}{3}
\end{aligned}\right\} \qquad (9.10)$$

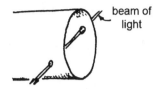

Fig. 9.16 Tracer measurement through the pipe centerline

(d) Response curves taken with line measurements
So far we have considered planar and flux measurements at the vessel exit. However, tracer concentration is often measured along a line (usually going through the center of the pipe) (Fig. 9.16).

For this type of measurement, the **E** and **F** curves and their various relationships are

In general

$$\mathbf{E}_\theta^{*\ell} = \theta \mathbf{E}^\ell, \quad \mathbf{E}_\theta^\ell = \frac{d\mathbf{F}^\ell}{d\theta}, \quad \mathbf{E}_\theta^{*\ell} = \frac{d\mathbf{F}^{*\ell}}{d\theta}. \tag{9.11}$$

For a Newtonian in a circular pipe

$$\left.\begin{array}{c} \mathbf{E}_\theta^\ell = \dfrac{1}{4\theta^2}\sqrt{1 - \dfrac{1}{2\theta}} \quad \text{with } \theta > \dfrac{1}{2} \\[2mm] \mathbf{F}^\ell = \sqrt{1 - \dfrac{1}{2\theta}} \\[2mm] \mathbf{F}^{*\ell} = \dfrac{1}{4}\ell n\left\{2\theta\left[1 + \sqrt{1 - \dfrac{1}{2\theta}}\right]^2\right\} \end{array}\right\}. \tag{9.12}$$

For a power law fluid in a circular pipe

$$\left.\begin{array}{c} \mathbf{E}_\theta^\ell = \dfrac{n}{(3n+1)\theta^2}\left[1 - \dfrac{\theta_o}{\theta}\right]^{-1/n+1} \quad \text{with } \theta > \theta_o = \dfrac{n+1}{3n+1} \\[3mm] \mathbf{F}^\ell = \left(1 - \dfrac{\theta_o}{\theta}\right)^{n/n+1} \end{array}\right\}. \tag{9.13}$$

For a Bingham plastic in a circular pipe

$$\left.\begin{array}{c} \mathbf{E}_\theta^\ell = \dfrac{(1-m)\theta_o}{2\theta^2\sqrt{1 - (\theta_o/\theta)}} \quad \text{with } \theta \geq \theta_o = \dfrac{m^2 + 2m + 3}{6}, \quad m = \dfrac{r_{plug}}{R} \\[3mm] \mathbf{F}^\ell = m + (1-m)\sqrt{1 - \dfrac{\theta_o}{\theta}} \end{array}\right\}. \tag{9.14}$$

For falling films or flow between parallel plates, with Newtonians

$$\mathbf{E}^\ell = \mathbf{E}^*, \quad \mathbf{E}^{*\ell} = \mathbf{E}^{**}, \quad \mathbf{F}^\ell = \mathbf{F}^*, \quad \mathbf{F}^{*\ell} = \mathbf{F}^{**}. \tag{9.15}$$

Problem

1. We want to model the flow of fluid in a circular flow channel. For this we locate three measuring points A, B, and C, 100 m apart along the flow channel. We inject tracer upstream of point A, fluid flows past points A, B, and C with the following results:

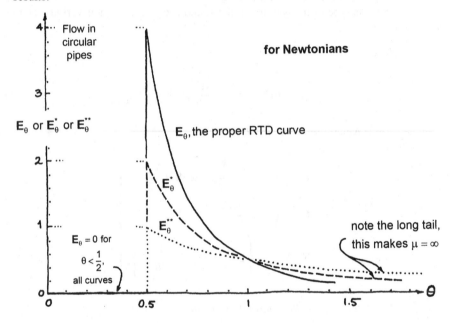

At A the tracer width is 2 m
At B the tracer width is 10 m
At C the tracer width is 14 m

What type of flow model would you try to use to represent this flow: dispersion, convective, tanks in series, or none of these? Give a reason for your answer.

References

Wen and Fan: **E, E*, F, F*** for power law fluids
Kim and Harris CES 28 1653 (1973): line measurements of Newtonians
Dalkeun Park, PhD thesis Oregon State University 1980: equations for **E****, **F**** and for Bingham plastics.
Lin CES 35 1477 (1980): power law fluids in annuli
Ananthakrishnan et al., AIChE J 11 1063 (1965)

Chapter 10
Batch Systems

In a batch system, fluid does not make only one pass past a measuring point in the system, it circulates and comes back again and again to the tracer measuring point. Thus, a pulse of tracer is seen again and again as it eventually becomes uniform in the system.

Let the circulation time through the vessel be t_{circ} and let the time for tracer to spread past the measurement location be t_{pass}. Then, how we treat these batch systems depends on the relative values of t_{circ} and t_{pass}. Figure 10.1 shows this.

10.1 Regime (a) $t_{pass} \ll t_{circ}$

With no overlap of the first and second pass, this case is easy to analyze. Just consider the first pass as part of a flow system and use the tanks-in-series model or the dispersion model. Example 1 illustrates this approach.

10.2 Regime (b) $t_{pass} < t_{circ}$

When there is a small time overlap, the treatment is still rather simple and is best done with the tanks-in-series model.

We look at the vessel as an N tanks-in-series system. We introduce a pulse signal into this N stage system, as shown in Fig. 10.2, the recorder will measure tracer as it flows by the first time.

O. Levenspiel, *Tracer Technology*, Fluid Mechanics and Its Applications 96,
DOI 10.1007/978-1-4419-8074-8_10, © Springer Science+Business Media, LLC 2012

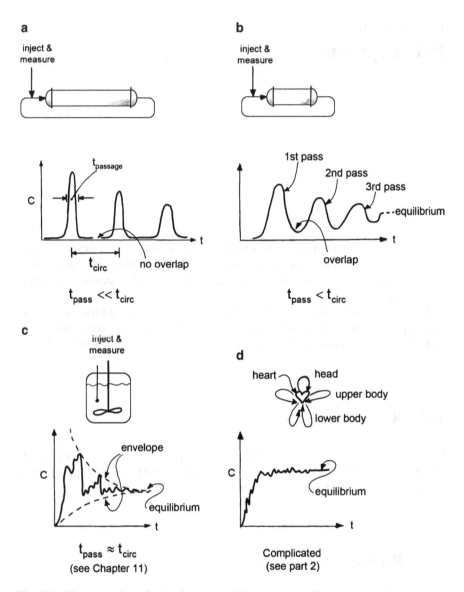

Fig. 10.1 Various regimes for batch systems. (**a**) $t_{pass} \ll t_{circ}$, (**b**) $t_{pass} < t_{circ}$, (**c**) $t_{pass} \approx t_{circ}$ (see Chap. 11), (**d**) complicated)

To obtain the output signal for these systems, simply sum up the contributions from the first, second, and succeeding passes. If m is the number of passes and n is the number of tanks-in series, we then have from (8.1).

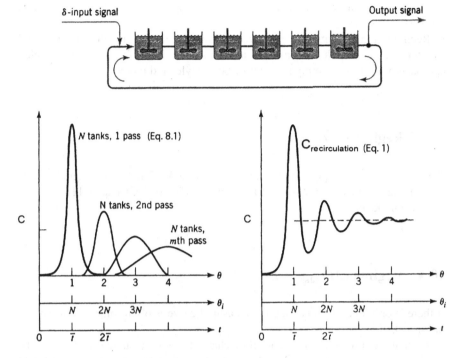

Fig. 10.2 Tracer signal in a recirculating system

$$\bar{t}_i C_{\text{pass}} = \left(e^{-t/\bar{t}_i}\right) \sum_{m=1}^{\infty} \frac{(t/\bar{t}_i)^{mN-1}}{(mN-1)!}, \tag{10.1a}$$

$$C_{\theta i,\text{pass}} = \left(e^{-\theta_i}\right) \sum_{m=1}^{\infty} \frac{\theta_i^{mN-1}}{(mN-1)!}, \tag{10.1b}$$

$$C_{\theta,\text{pass}} = \left(Ne^{-N\theta}\right) \sum_{m=1}^{\infty} \frac{(N\theta)^{mN-1}}{(mN-1)!}. \tag{10.1c}$$

As an example of the expanded form of (10.1) we have for five tanks-in-series

$$C_{\text{pass}} = \frac{5}{\bar{t}} e^{-5t/\bar{t}_i} \left[\frac{(5t/\bar{t})^4}{4!} + \frac{(5t/\bar{t})^9}{9!} + \cdots \right], \tag{10.2a}$$

$$C_{\theta i,\text{pass}} = e^{-\theta_i} \left[\frac{\theta_i^4}{4!} + \frac{\theta_i^9}{9!} + \frac{\theta_i^{14}}{14!} + \cdots \right], \tag{10.2b}$$

$$C_{\theta,\text{pass}} = 5e^{-5\theta} \left[\frac{(5\theta)^4}{4!} + \frac{(5\theta)^9}{9!} + \cdots \right], \tag{10.2c}$$

where the terms in brackets represent the tracer signal from the first, second, and successive passes.

Recirculation systems can be represented equally well by the dispersion model (see van der Vusse 1962; Vonken et al. 1964; Harrell and Perona 1968). Which approach one takes is simply a matter of taste, style, and mood.

10.3   Regime (c) $t_{pass} \approx t_{circ}$

This regime often represents a batch stirred tank. The output signal seems randomlike but lies in an envelope as shown in Fig. 10.1c. Chapter 11 considers this case in detail.

10.4   Regime (d) $t_{pass} \geq t_{circ}$

If there is only one flow channel, the measured curve is often somewhat as shown in Fig. 10.1d.

If there are a number of parallel flow channels, as in an animal body, then the response is quite complicated with overlapping signals. In this situation, it is advisable to trace the signal in one part of the body and then in another part of the body.

Figure 10.3 shows an example of such a system.

10.5   Rapid Recirculation with Throughflow

Here is a situation which has characteristics of both flow and batch systems. For relatively rapid recirculation compared to throughflow, the system as a whole acts as one large stirred tank; hence, the observed tracer signal is simply the superposition of the recirculation pattern and the exponential decay of an ideal stirred tank. This is shown in Fig. 10.4 where C_0 is the concentration of tracer if it is evenly distributed in the system.

This form of curve is encountered in closed recirculation systems in which tracer is broken down and removed by a first-order process, or in systems using radio-active tracers. Drug injection on living organisms gives this sort of superposition because the drug is constantly being slowly eliminated by the organism.

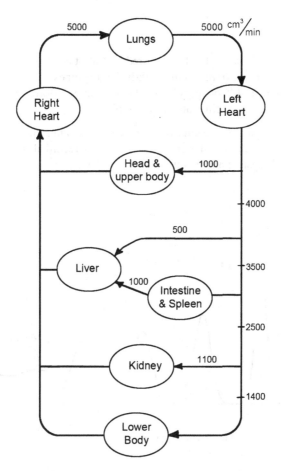

Fig. 10.3 Simplified blood circulation in a human. Bischoff and Brown (1966)

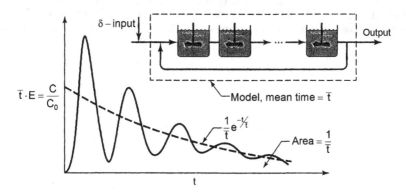

Fig. 10.4 Recirculation with slow throughflow

Problems

1. Strong radioactive waste fluids are stored in "safe-tanks" which are simply long small-diameter (e.g., 20 m by cm) slightly sloping pipes. To avoid sedimentation and development of "hot spots," and also to insure uniformity before sampling the contents, fluid is recirculated in these pipes.

   To model the flow in these tanks, fluid is recirculated in a closed loop, a pulse of tracer is introduced, and the curve (Fig. 10.5) is recorded. Develop a suitable model for this system and evaluate its parameters.

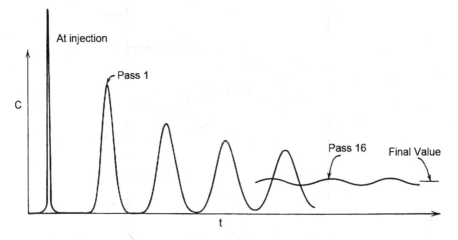

Fig. 10.5 RTD for a closed recirculating system

References

Van der Vusse, J.G., *Chem. Eng. Sci.*, 17 507 (1962).
Vonken, R.M., Holmes, D.B. and den Hartog, H.W., *Chem. Eng. Sci.*, 19 209 (1964).
Harrell, J.E., Jr. and Perona, J.J., *Ind. Eng. Chem. Proc. Des. Develop.*, 7 464 (1968).
Bischoff, K.B. and Brown, R.G., *Drug Distribution in Mammals* CEP Symp. Ser. pg. 333, No 66 62 1966.

Chapter 11
The Stirred Tank: Mixing Time and Power Requirement

When using a rotating impeller to mix fluids contained in a tank, we ask two questions:

- What size motor do we need for a given impeller rotation speed?
- How long must we mix the fluid to achieve a required degree of uniformity?

The well-known correlation between power number, N_p, and Reynolds number, N_{Re} (Rushton et al. 1950; Bates et al., 1963; Uhl and Gray 1966; Holland and Chapman 1966; Sterbacek and Tausk 1965)) helps us with the first question. A mixing-rate number (Khang 1975; Khang and Levenspiel 1976) to characterize the rate of approach to uniformity is obtained from tracer methods and shows how to answer the second question. We then extend the N_p/N_{Re} chart to include the mixing-rate number.

Residence-time distribution theory and statistical considerations (Khang 1975; Khang and Levenspiel 1976) predict that any nonhomogeneity, as measured by the concentration fluctuations about the final value, A, should decay with time exponentially:

$$A = 2e^{-Kt} \quad \text{or} \quad \ell n(A/2) = -Kt, \tag{11.1}$$

where K is the decay constant. For example, if a quantity of foreign fluid is added to the mixing tank, it should blend as shown in Fig. 11.1. The power needed for such an operation is then found from $\mathbf{K}(s^{-1})$, $\mathbf{n}$(rev/s or RPM) and the derived expressions of Khang and Levenspiel.

For large enough Reynolds numbers ($N_{Re} > 10^4$), these charts reduce to the simple relationships:

For turbine mixers:

$$\frac{n}{K}\left(\frac{d}{D}\right)^{2.3} = 0.1\left(\frac{Pg_c}{\rho n^3 d^5}\right) = 0.5. \tag{11.2}$$

O. Levenspiel, *Tracer Technology*, Fluid Mechanics and Its Applications 96,
DOI 10.1007/978-1-4419-8074-8_11, © Springer Science+Business Media, LLC 2012

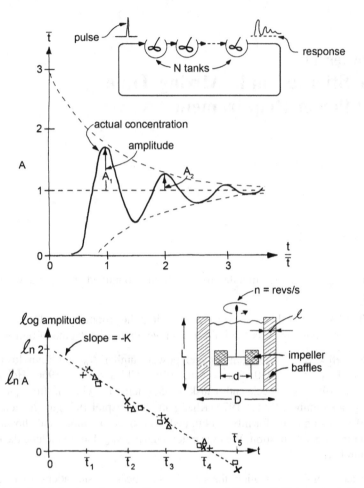

Fig. 11.1 Typical behavior when blending a quantity of foreign fluid into the bulk liquid contained in a baffled mixing tank

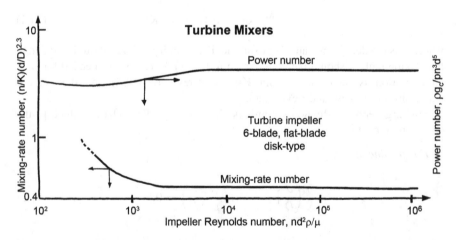

Fig. 11.2 Dimensionless correlations for turbine mixers. Speed, scaleup, and design procedures for agitated system, from Khang and Levenspiel (1976)

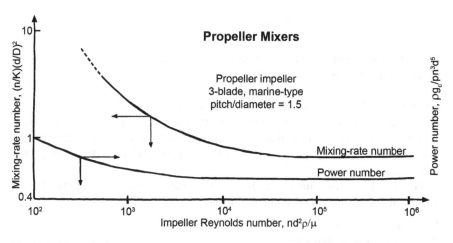

Fig. 11.3 Dimensionless correlations for propeller mixers. Speed, scaleup, and design procedures for agitated systems, from Khang and Levenspiel (1976)

For propeller mixers:

$$\frac{n}{K}\left(\frac{d}{D}\right)^2 = 1.5\left(\frac{Pg_c}{\rho n^3 d^5}\right) = 0.9. \tag{11.3}$$

One direct consequence of (11.2) and (11.3) is that the power required to achieve any degree of uniformity in a fixed time depends on impeller size. Hence:

For turbine mixers:	$P \propto d^{1.9}$
For propeller mixers:	$P \propto d^{1}$

Thus, larger impellers need less power for a given job.

These correlations apply also to fully baffled tanks with vertical impellers at the center. For other tank geometries somewhat different curves can be expected. All these graphs and equations relate:

The time of mixing (t)
Power requirement (P)
Impeller size (d)
Tank size (D)
Stirrer speed (n) rev/s (s^{-1})

needed to achieve any required degree of uniformity of the mixture. The charts and equations should be useful for design and scaleup of these agitation systems.

11.1   Applying the Correlations

The following examples illustrate some of the many types of problems that may be treated with (11.1) and (11.2), or (11.1) and (11.3).

Example 1. *How Long to Blend Away Nonuniformities.*
A small batch of salt solution is introduced into the water in a mixing tank. How long will it take to blend away the salt solution so that the concentration fluctuations about the final value are not greater than 0.1% of original and what is the power needed for this operation?

The impeller is a six flat blade, disc-type turbine, positioned in the middle of a fully baffled tank. Here is additional information:

$D = 6$ m, tank diameter
$d = 1.2$ m, impeller diameter
$L = 6$ m, liquid depth
$\rho = 1,000$ kg/m^3, fluid density
$\mu = 0.001$ kg/m·s, fluid viscosity
$n = 0.2$ rev/s; 12 RPM, impeller rotation rate

Solution. First calculate the impeller Reynolds number

$$Re = \frac{nd^2\rho}{\mu} = \frac{0.2(1.2)^2 1,000}{0.001} = 2.9 \times 10^5$$

Since $Re > 10^4$, (11.2) gives

$$K = \frac{n}{0.5}\left(\frac{d}{D}\right)^{2.3} = \frac{0.2}{0.5}\left(\frac{1.2}{6}\right)^{2.3} = 0.01 \text{ s}^{-1}$$

Then from (11.1)

$$t = \frac{1}{-K}\ell n\left(\frac{A}{2}\right) = \frac{1}{-0.01}\ell n\left(\frac{0.001}{2}\right) = 760 \text{ s} \approx 13 \text{ min}$$

The power needed, from (11.2), is

$$P = \frac{5\rho n^3 d^5}{g_c} = \frac{5(1,000)(0.2)^3(1.2)^5}{1} = \underline{\underline{100 \text{ W}}}$$

Example 2. *How Much Power Is Needed to Reduce the Mixing Time?.*
For the equipment of Example 1 find the power needed to reduce the mixing time from 13 to 5 min.

Solution. First find the decay rate constant from (11.1)

$$K = \frac{-1}{t} \ln\left(\frac{A}{2}\right) = \frac{-1}{300} \ln\left(\frac{0.001}{2}\right) = 0.02534 \text{ s}^{-1}$$

Now the rotational speed of the impeller, from (11.2), is

$$n = 0.5K\left(\frac{D}{d}\right)^{2.3} = 0.5(0.02534)\left(\frac{6}{1.2}\right)^{2.3} = 0.513 \text{ rev/s} \cong 31 \text{ RPM.}$$

And again from (11.2)

$$P = \frac{5\rho n^3 d^5}{g_c} = \frac{5(1,000)(0.513)^3(1.2)^5}{1} = \underline{\underline{1,680 \text{ W.}}}$$

Example 3. *Impeller Size Needed to Reduce Mixing Time.*
For a 5 min mixing time, what diameter and RPM of impeller will give the minimum power consumption? Note: the largest allowable impeller diameter $d = D/2$.

Solution. First, choose the largest allowable impeller diameter, $d = 6/2 = 3$ m. Then

$$K = \frac{-1}{t} \ln\left(\frac{A}{2}\right) = \frac{-1}{300} \ln\left(\frac{0.001}{2}\right) = 0.02534 \text{ s}^{-1}$$

Noting from Example 1 that Re $> 10^4$, (11.2) gives

$$n = 0.5K\left(\frac{D}{d}\right)^{2.3} = 0.5(0.02534)\left(\frac{6}{3}\right)^{2.3} = 0.0624 \text{ rev/s} = 3.7 \text{ RPM.}$$

Then the power needed, from (11.2), is

$$P = \frac{0.5}{0.1} \frac{\rho n^3 d^5}{g_c} = 5\left(\frac{(1,000)(0.0624)^3(3)^5}{1}\right) = \underline{\underline{294 \text{ W.}}}$$

To summarize

	Mixing time needed (min)	Rotation rate (RPM)	Power needed (W)
Example 1 Small impeller Long-run time	~13	12	100
Example 2 Small impeller Short-run time	5	31	1,680
Example 3 Large impeller Short-run time	5	3.7	294

Nomenclature

A	Normalized amplitude of concentration fluctuations about the final value (see Fig. 11.1), dimensionless
d	Diameter of impeller, m
D	Diameter of tank, m
g_c	Newton's law conversion factor, $1 \text{ N s}^2/\text{kg·m}$
K	Amplitude decay-rate constant, s
l	Baffle width, m
L	Liquid depth, m
n	Impeller rotational rate, rev/s, (s^{-1})
N_{Re}	Impeller Reynolds number, dimensionless
P	Mixing power, W
t	Time, s
μ	Viscosity, kg/m s
ρ	Density, kg/m^3

Problems

1. Consider Example 1. With the small impeller it took a 100 W motor 13 min to obtain the desired degree of mixing. Suppose we use a larger impeller ($d = 3$ m). What mixing time and power would be needed?

2. Fructose is very useful in today's kitchens; however, it is a colorless viscous fluid which is not very attractive.

I'm thinking of making a market study of an attractively colored fructose which I will call "fruityose." For a test I will put a slug of pink dye into a $D = 1$ m, $L = 1$ m fructose-filled vat and mix it with a propeller type stirrer, $d = 0.33$ m.

How long will we have to mix to get color fluctuations not more than 0.1% of the equilibrium?

Data   For Karo syrup (fructose): $\rho = 1,300 \text{ kg/m}^3$
$$\mu = 180 \times 10^{-3} \text{ kg/m s}$$

3. There are too many cases of children drinking lemon colored lemon flavored poisonous cleaning fluid A. To discourage this, we plan to introduce a small amount of foul tasting material B into the vat containing A, $D = 2$ m and $L = 2$ m, using a 6 blade turbine mixer, $d = 0.5$ m.

How long should we mix the ingredients to get a close to uniform mixture (0.01% deviation from uniformity)?

Data   Property of the cleaning fluid: $\rho = 980 \ \text{kg/m}^3$
$\mu = 1.2 \times 10^{-3} \ \text{kg/m s}$

References

Rushton, J.H., and others, *Chem. Eng. Progr.*, 46, 395, 467 (1950).

Bates, R.L., Fondy, P.L., and Corpstein, R.R., *Ind. Eng. Chem. Process Design Develop.*, 2, 310 (1963).

Uhl, V.W. and Gray, J.B., "Mixing," Vol. I, Academic Press, New York (1966).

Holland, F.A. and Chapman F.S., "Liquid Mixing and Processing in Stirred Tanks," Reinhold, New York (1966).

Sterbacek, Z. and Tausk, P., "Mixing in the Chemical Industry," International Series of Monographs in Chemical Engineering, Vol. 5, Pergamon, London (1965).

Khang, S.J., Ph.D. thesis, Oregon State University, Corvallis, OR (1975).

Khang, S.J. and Levenspiel, O., *Chem. Eng. Sci.*, 31, 569–577 (1976).

Chapter 12
Meandering Flow and Lateral Dispersion

Until this point we have considered a process vessel with flow entering at one point and flow leaving at a second point, all at a steady rate:
In the vessel the fluid can ideally flow in a variety of ways

- Plug flow
- Mixed flow
- Through a number of ideal mixed flow regions in series
- At the vessel entrance and exit we only have plug flow (Fig. 12.1)

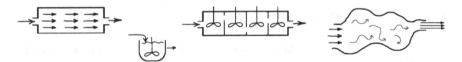

Fig. 12.1 Ideal flow models

In real vessels the flow approximates these ideals. Tracer technology can tell how the process, the reactor or the mass transfer unit proceeds in such systems.

In situations where fluid may enter at more than one point and may leave at more than one point, for example see Fig. 12.2, we have a much more complicated situation to treat and the simpler methods of this book can not be used.

In this chapter we consider just one such complicated situation – where fluid passing through the vessel experiences two types of mixing, *meandering flow* and *lateral dispersion*.

Consider flow in one direction with an input of tracer at one point, say $z = 0$, for example air flow past a smoking chimney. Then two phenomena cause the smoke to spread into its surroundings: first the eddy type dispersion such as the diffusion of smoke "d"; and second the swirling and bulk flow of smoke which we call meandering, "m."

O. Levenspiel, *Tracer Technology*, Fluid Mechanics and Its Applications 96,
DOI 10.1007/978-1-4419-8074-8_12, © Springer Science+Business Media, LLC 2012

Fig. 12.2   Two entering or leaving streams; not treated here

Now *without meander* the tracer will flow as shown in Fig. 12.3

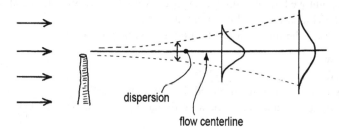

Fig. 12.3   Flow with lateral dispersion

But *with meander* we would get a somewhat more chaotic situation as shown in Fig. 12.4

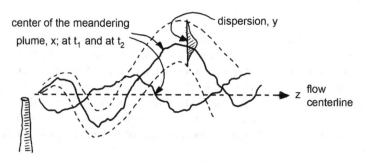

Fig. 12.4   Flow with meander and dispersion

We show these two lateral movements with their Gaussian distributions for a typical curve in Fig. 12.5. A mathematical model of this behavior was first created by Wilson (1904) and some 50 years later by Carslaw and Jeager (1959) and then by Jovanovic et al. (1980). A simplified treatment was recently proposed by Fitzgerald (Personal communication, 2011) and is presented here.

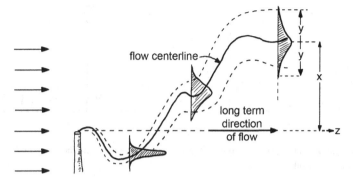

Fig. 12.5 Lateral meander and dispersion of a tracer plume

Let z = the distance downstream from the introduction of tracer.
  x = the bulk lateral movement of tracer from its flow centerline, the meander effect
  y = spread of tracer from its flow centerline, the dispersion effect

Then, to account for this chaotic behavior, assume that the small-scale eddy and molecular diffusion produces a Normal distribution in space, with a variance of $\mathbf{V_d}$, and that the meandering plume also flickers with a Normal distribution in space, $\mathbf{V_m}$. When dispersion and meander both occur then the tracer concentration measured at any downstream location z is also Normal, but with a variance which is the sum of the diffusion and meander variances, or

$$\mathbf{V}[C] = \mathbf{V}[C_d] + \mathbf{V}[C_m]. \tag{12.1}$$

If we square the measured signals with time at lateral points y_i and at the downstream location z, we then get another Normal curve whose dispersion variance is one-half of the original dispersion variance. Thus the squared concentration values have a variance which is the sum of the meander plus one-half of the original dispersion variance or

$$\mathbf{V}[C^2] = \mathbf{V}[C_m] + \frac{1}{2} \cdot \mathbf{V}[C_d]. \tag{12.2}$$

Experimentally, we first record the tracer concentration C at three lateral positions $i = A, B, D$, or maybe more, all at the same distance downstream z, from the tracer inlet which is at $z = 0$, and record these concentrations for a rather long period of time, say from t_1 to t_2. Also determine the average concentration in this time interval, shown in Fig. 12.6

$$\overline{C_i} = \int_{t_1}^{t_2} \frac{C_i dt}{t_2 - t_1} \quad i = A, B, D. \tag{12.3}$$

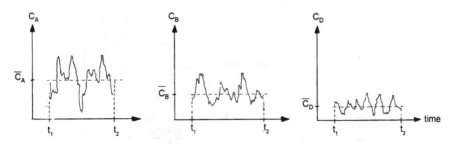

Fig. 12.6 Plot the tracer reading at points A, B, D between t_1 and t_2 and their averages $\overline{C_A}, \overline{C_B}$ and $\overline{C_D}$ in this time period t_1 to t_2

Also plot the squares of C_A, C_B, C_D vs. time and then find and plot their three averages as shown in Fig. 12.7

$$\overline{C_i^2} = \int_{t_1}^{t_2} \frac{D_i^C \, dt}{t_2 - t_1} \quad i = A, B, D. \tag{12.4}$$

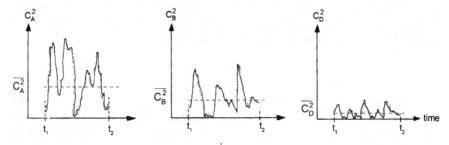

Fig. 12.7 Plot the tracer reading C_A^2 at points A,B,D between t_1 and t_2 and their averages, $\overline{C_A^2}, \overline{C_B^2}$ and $\overline{C_D^2}$ in this time period t_1 to t_2

According to Jovanovic et al. (1980) and Fitzgerald (Personal communication, 2011) these lateral spreads are Gaussian so they are additive with time. Using this property, shown in (12.1) and (12.2), we have

$$\mathbf{V}[C_d] = 2\mathbf{V}[C] - 2\mathbf{V}[C^2], \tag{12.5}$$

$$\mathbf{V}[C_m] = -\mathbf{V}[C] + 2\mathbf{V}[C^2]. \tag{12.6}$$

Relating the variances to their dispersion coefficients

$$\mathbf{V}[C_d] = \mathbf{D}_d \cdot (z/u), \tag{12.7}$$

$$\mathbf{V}[C_m] = \mathbf{D}_m \cdot (z/u). \tag{12.8}$$

Therefore the dispersion coefficients are

$$\mathbf{D}_d = \mathbf{V}[C_d] \cdot (u/z), \tag{12.9}$$

$$\mathbf{D}_m = \mathbf{V}[C_m] \cdot (u/z). \tag{12.10}$$

Now what can be the use of all this? This approach was originally suggested by Fitzgerald and Jovanovic (1979) for evaluation of gas mixing processes in fluidized beds. Gifford (1950) also discussed a general approach in analyzing statistical properties of fluctuating plumes.

For example, to predict or account for the progress of a chemical reaction between two entering fluids in a *vessel without internals* it is often $\mathbf{D}_d$ that tells how much mixing and reaction occurs, while $\mathbf{D}_m$ hardly contributes to the mixing of fluids. However, in some situations the opposite occurs. It all depends on the geometry of the vessel. Thus:

- If the vessel contains a tube bank that could change the mixing characteristics of the vessel.
- In packed beds having two separate reactant streams, $\mathbf{D}_m$ plays hardly any role in the mixing of the two reactant streams.
- In a bubbling fluidized bed with big bursting bubbles the meandering phenomenon with its $\mathbf{D}_m$ dominates the mixing; however at a low gas velocity just a bit above minimum fluidization $\mathbf{D}_d$ enters the picture.
- In G/G, L/L and G/S reactions it is very important to know which factor dominates, because here the mixing phenomenon controls the progress of the reaction.

In summary, designers should focus and measure the role of $\mathbf{D}_m$ and $\mathbf{D}_d$ for G/G, L/L and for G/S reactions for various bed geometries.

Example 1. *Horizontal Spread of Tracer in a Large Fluidized Bed.*
Add a steady stream of tracer gas into the center bottom of a 1 m diameter, 1 m high gently fluidized bed, and sample the rising gas at four points, all at $z = 80$ cm above the bottom;

Point $A[-y_A] = -1.5$ cm, due East of the vessel center
Point $B - y_B = -0.5$ cm due East
Point $D - y_D = 0.5$ cm due West
Point $E - y_E = 1.5$ cm due West

The upflow of fluidizing gas and tracer in the bed is $u = 20$ cm/s

The normalized concentration of a tracer at these locations at 19 1-min time intervals and then their mean values are shown in the table below

	$A=-1.5$	$B=-0.5$	$D=0.5$	$E=1.5$	$A=-1.5$	$B=-0.5$	$D=0.5$	$E=1.5$
t (min)	$C(t)$ at A	$C(t)$ at B	$C(t)$ at D	$C(t)$ at E	$C^2(t)$ at A	$C^2(t)$ at B	$C^2(t)$ at D	$C^2(t)$ at E
1	0.07	0.66	0.88	0.16	0.00	0.43	0.78	0.03
2	0.79	0.77	0.10	0.00	0.62	0.59	0.01	0.00
3	0.34	1.00	0.40	0.02	0.11	1.00	0.16	0.00
4	0.00	0.12	0.81	0.74	0.00	0.01	0.66	0.55
5	0.01	0.21	0.94	0.56	0.00	0.05	0.89	0.32
6	0.00	0.07	0.69	0.86	0.00	0.01	0.47	0.74
8	0.37	1.00	0.37	0.02	0.13	1.00	0.14	0.00
9	0.09	0.75	0.81	0.12	0.01	0.56	0.65	0.01
10	0.06	0.65	0.89	0.16	0.00	0.42	0.79	0.03
11	1.00	0.36	0.02	0.00	1.00	0.13	0.00	0.00
12	0.03	0.47	0.98	0.28	0.00	0.22	0.97	0.08
13	0.05	0.60	0.92	0.19	0.00	0.37	0.84	0.04
14	0.00	0.10	0.76	0.79	0.00	0.01	0.58	0.63
15	0.40	1.00	0.34	0.02	0.16	1.00	0.12	0.00
16	0.03	0.48	0.98	0.27	0.00	0.23	0.96	0.07
17	0.10	0.77	0.79	0.11	0.01	0.59	0.62	0.01
18	0.63	0.90	0.18	0.00	0.40	0.82	0.03	0.00
19	0.30	0.99	0.44	0.03	0.09	0.98	0.20	0.00
$\overline{C_i}=$	0.23	0.61	0.64	0.23	$\overline{C_i^2}=0.13$	0.47	0.50	0.13

Estimate the variance and dispersion coefficients, for lateral dispersion and for meandering of the fluidizing gas at A, B, D, E.

For the distribution of concentration readings

$$\mathbf{V}[C] = \frac{\sum y_i^2 \overline{C_i}}{\sum \overline{C_i}} - \left[\frac{\sum y_i \overline{C_i}}{\sum \overline{C_i}}\right]^2. \tag{12.11}$$

And for the concentration squared

$$\mathbf{V}[C^2] = \frac{\sum y_i^2 \overline{C_i^2}}{\sum \overline{C_i^2}} - \left[\frac{\sum y_i \overline{C_i^2}}{\sum \overline{C_i^2}}\right]^2. \tag{12.12}$$

For all four points, A, B, D, E, the sum of the means of their variances

$$\mathbf{V}[C] = \left[\begin{array}{l} +\dfrac{2.25(0.23) + 0.25(0.61) + 0.25(0.64) + 2.25(0.23)}{0.23 + 0.61 + 0.64 + 0.23} \\[2ex] -\left[\dfrac{-1.5(0.23) - 0.5(0.61) + 0.5(0.64) + 1.5(0.23)}{1.71}\right]^2 \end{array} \right] \tag{12.13}$$

$$= 0.7854$$

Similarly, for the squares of their variances

$$\mathbf{V}[C^2] = \begin{bmatrix} +\dfrac{2.25(0.13) + 0.25(0.47) + 0.25(0.50) + 2.25(0.13)}{0.13 + 0.47 + 0.50 + 0.13} \\ -\left[\dfrac{-1.5(0.13) - 0.5(0.47) + 0.5(0.50) + 1.5(0.13)}{1.23}\right]^2 \end{bmatrix} \quad (12.14)$$

$$= 0.6728 - 0.0001 = 0.6727.$$

$$\mathbf{V}[C_d] = 2(0.7854) - 2(0.6727) = 0.2254 \text{ cm}^2 \qquad (12.15)$$

$$\mathbf{V}[C_m] = 2(0.6727) - 2(0.7854) = 0.56 \text{ cm}^2 \qquad (12.16)$$

$$\mathbf{D}_d = \mathbf{V}[C_d] \cdot \frac{u}{z} = 0.2254 \text{ cm}^2 \left(\frac{20}{80}\right) = 0.56 \frac{\text{cm}^2}{\text{s}} \qquad (12.17)$$

$$\mathbf{D}_m = \mathbf{V}[C_m] \cdot \frac{u}{z} = 0.56 \left(\frac{20}{80}\right) = 0.14 \frac{\text{cm}^2}{\text{s}} \qquad (12.18)$$

Notation

$$\mathbf{V}[C] = \sigma^2(C),$$

$$\mathbf{V}[C^2] = \sigma^2(C^2) = \sigma^2(C \cdot C).$$

References

1. "The Mixing of Tracer Gas in Fluidized Beds of Large Particles," G.N. Jovanovic, N. Catipovic, T. Fitzgerald and O. Levenspiel, in *Fluidization* (Grace and Matsen – Editors), Plenum Press, New York, 325 (1980).
2. T. Fitzgerald and G. N. Jovanovic, *Gas Flow in Fluidized Beds of Large Particles, Experiment and Theory*, PhD Project, Oregon State University, Chemical Engineering Department, 1979.
3. F. Gifford, "Statistical Properties of a Fluctuating Plume Dispersion Model," *Advances in Geophysics*, Academic Press Vol 6; 199 (1950)
4. H. A. Wilson, "On Convection of Heat," Proc. Roy. Soc., 12, 408 February 15, 1904, London
5. H.S. Carslaw, and J. C. Jeager "Conduction of Heat in Solids," Oxford at the Clarendon Press (1959).
6. This chapter was prepared by Tom Fitzgerald.TomdFitz@gmail.com

Chapter 2
The Mean and Variance of a Tracer Curve

O. Levenspiel, *Tracer Technology*, Fluid Mechanics and Its Applications 96,
DOI 10.1007/978-1-4419-8074-8, © Springer Science+Business Media, LLC 2012

DOI 10.1007/978-1-4419-8074-8_13

Page 7, equation 2.6
Please insert = and change last = to a − (minus)

$$\sigma^2 = \frac{\sum t_i^2 C_i \Delta t_i}{\sum C_i \Delta t_i} - \bar{t}^2 \ \frac{for\ equal}{\Delta t_i} = \frac{\sum t_i^2 C_i}{\sum C_i} - \bar{t}^2.$$

Page 8
Please insert an 'A' at the beginning of the sentence.

2. For symmetrical S-shaped data
   A smooth S-shaped step-response curve often corresponds to …

The online version of the original chapter can be found at
http://dx.doi.org/10.1007/978-1-4419-8074-8_2

Chapter 3
The E and E_θ Curves from Pulse and Step Tracer Experiments

O. Levenspiel, *Tracer Technology*, Fluid Mechanics and Its Applications 96,
DOI 10.1007/978-1-4419-8074-8, © Springer Science+Business Media, LLC 2012

DOI 10.1007/978-1-4419-8074-8_13

Page 21
Fig. 3.15 F value from the last of Table EZ should be Fig. 3.15 F value from Table E2.

Page 22
On line 3 - Please change this equation from $\dfrac{\mathrm{m}}{\mathrm{v}}$ to $\dfrac{\dot{\mathrm{m}}}{v}$

(where you remove the line under the 'm' and add a dot over the 'm' and italicize the 'v')

Page 22
On line 4 - Please insert 'However' before *from the above graph* and add a ':' after *graph*

  However from the above graph:

The online version of the original chapter can be found at
http://dx.doi.org/10.1007/978-1-4419-8074-8_3

ERRATUM

Chapter 6
The Mean and Variance of a Tracer Curve

O. Levenspiel, *Tracer Technology*, Fluid Mechanics and Its Applications 96,
DOI 10.1007/978-1-4419-8074-8, © Springer Science+Business Media, LLC 2012

DOI 10.1007/978-1-4419-8074-8_13

Page 52, equation 6.4

First row of equation: Please change the t in the denominator to a t with a line above it

$$= \frac{\bar{t}_E}{t} = \frac{\bar{t}_E v}{V} = 1$$

to

$$= \frac{\bar{t}_E}{\bar{t}} = \frac{\bar{t}_E v}{V} = 1$$

The online version of the original chapter can be found at
http://dx.doi.org/10.1007/978-1-4419-8074-8_6

Chapter 7
Intermixing of Flowing Fluids

O. Levenspiel, *Tracer Technology*, Fluid Mechanics and Its Applications 96,
DOI 10.1007/978-1-4419-8074-8, © Springer Science+Business Media, LLC 2012

DOI 10.1007/978-1-4419-8074-8_13

Page 75
Please add (see Fig. 6.3) to the sentence on line 4
"Since $\mathbf{D}/ul < 0.01$, the tracer curve is symmetrical, (see Fig. 6.3) so"

The online version of the original chapter can be found at
http://dx.doi.org/10.1007/978-1-4419-8074-8_7

Chapter 8
The Tanks-in-Series Model

O. Levenspiel, *Tracer Technology*, Fluid Mechanics and Its Applications 96,
DOI 10.1007/978-1-4419-8074-8, © Springer Science+Business Media, LLC 2012

DOI 10.1007/978-1-4419-8074-8_13

Page 85
Figure 8.5, Please make the D in the equation a bold upright **D**

$$\sigma_\theta^2 = 2\left(\frac{D}{uL}\right) = \frac{1}{N}$$

to

$$\sigma_\theta^2 = 2\left(\frac{\mathbf{D}}{uL}\right) = \frac{1}{N}$$

Page 85
At the bottom of page 85, in the reference, please add CES **17** 576
"For large deviation from plug flow, see Levenspiel CES **17** 576 (1962)"

The online version of the original chapter can be found at
http://dx.doi.org/10.1007/978-1-4419-8074-8_8

Chapter 12
Meandering Flow and Lateral Dispersion

O. Levenspiel, *Tracer Technology*, Fluid Mechanics and Its Applications 96,
DOI 10.1007/978-1-4419-8074-8, © Springer Science+Business Media, LLC 2012

DOI 10.1007/978-1-4419-8074-8_13

Page 130
The equation 12.4 in page 130 is erroneous. We have corrected it to make it appear as follows:

$$\overline{C_i^2} = \int_{t_1}^{t_2} \frac{C_i^2 \mathrm{d}t}{t_2 - t_1} \quad i = A, B, D.$$

Page 131
In the last sentence tracaer is misspelled. It should be tracer.

Page 133
In equation 12.16 remove this '2'

$$\mathbf{V}[C_{\mathrm{m}}] = 2(0.6727) - 2(0.7854) = 0.56 \text{ cm}^2$$

to

$$\mathbf{V}[C_{\mathrm{m}}] = 2(0.6727) - (0.7854) = 0.56 \text{ cm}^2$$

The online version of the original chapter can be found at
http://dx.doi.org/10.1007/978-1-4419-8074-8_12

In equation 12.17 the last numeral should be $= 0.056$ NOT 0.56.

$$\mathbf{D_d} = \mathbf{V}[C_d] \cdot \frac{u}{z} = 0.2254 \text{ cm}^2 \left(\frac{20}{80}\right) = 0.56 \frac{\text{cm}^2}{\text{s}}$$

to

$$\mathbf{D_d} = \mathbf{V}[C_d] \cdot \frac{u}{z} = 0.2254 \text{ cm}^2 \left(\frac{20}{80}\right) = 0.056 \frac{\text{cm}^2}{\text{s}}$$

Index

O. Levenspiel, *Tracer Technology*, Fluid Mechanics and Its Applications 96,
DOI 10.1007/978-1-4419-8074-8, © Springer Science+Business Media, LLC 2012